Blowout

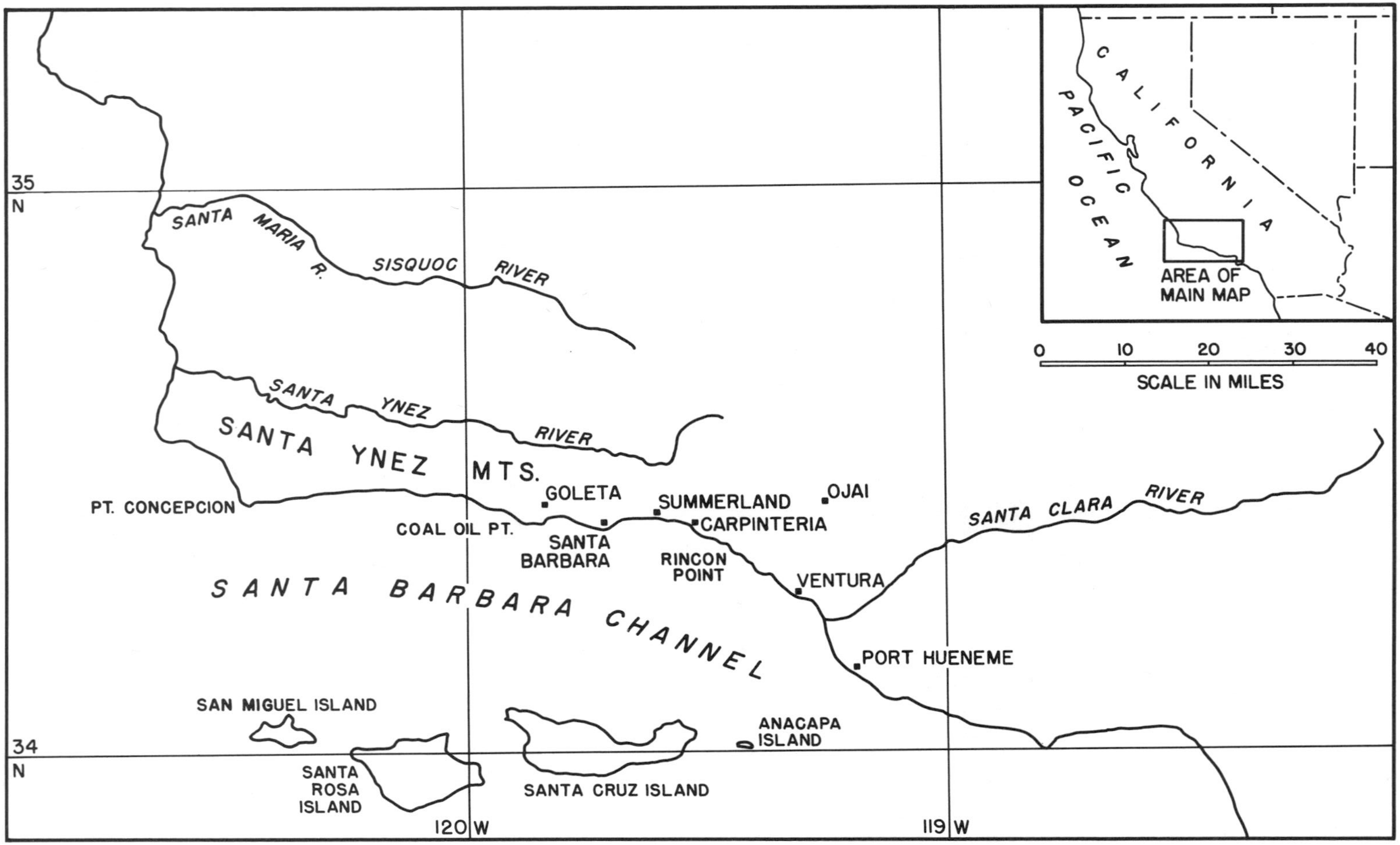

Santa Barbara Channel and surrounding area.

blowout

A Case Study of the Santa Barbara Oil Spill

by Carol E. Steinhart
and John S. Steinhart
UNIVERSITY OF WISCONSIN, MADISON

with a Foreword by Walter J. Hickel
FORMER SECRETARY OF THE INTERIOR

Duxbury Press
A DIVISION OF THE WADSWORTH PUBLISHING COMPANY, INC.
North Scituate, Massachusetts and Belmont, California

ACKNOWLEDGMENTS

The excerpts in this book are reprinted by permission from the following sources:

Auden, W. H.: From *Collected Shorter Poems 1927-1957* by W. H. Auden. Copyright 1949 by W. H. Auden. Reprinted by permission of Random House, Inc.

Brautigan, Richard: From *Rommel Drives On Deep Into Egypt* by Richard Brautigan. Copyright © 1970 by Richard Brautigan. A Seymour Lawrence Book/Delacorte Press. Reprinted by permission of the publisher.

Carson, Rachel: From *The Sea Around Us* by Rachel Carson, pp. 190-191, 1961. Reprinted by permission of Oxford University Press.

Frost, Robert: From *The Poetry of Robert Frost* edited by Edward Connery Latham. Copyright 1947, © 1969 by Holt, Rinehart and Winston, Inc. Reprinted by permission of Holt, Rinehart and Winston, Inc.

Laing, R. D.: From *Knots* by R. D. Laing. Copyright © 1970 by R. D. Laing. Reprinted by permission of Pantheon Books, a Division of Random House, Inc.

Simon, Paul: From "Keep the Customer Satisfied." © 1970 by Paul Simon. Used by permission of the publisher.

Sollen, Robert H.: From *Santa Barbara News-Press,* February 20, 1969. Reprinted by permission of the author.

Thurber, James: Copyright © 1956 by James Thurber. From *Further Fables for Our Time,* published by Simon & Schuster, New York. Originally printed in *The New Yorker.* Reprinted by permission of Mrs. James Thurber.

Tzu, Chuang: From Thomas Merton, *The Way of Chuang Tzu.* Copyright © 1965 by the Abbey of Gethsemani. Reprinted by permission of New Directions Publishing Corporation.

DUXBURY PRESS
A Division of Wadsworth Publishing Company, Inc.

L.C.Cat.Card No.: 72-075102
ISBN 0-87872-037-5
Printed in the United States of America

1 2 3 4 5 6 7 8 9 10 76 75 74 73 72

FOR EDITH MALIN

Foreword

Carol and John Steinhart's case study of the Santa Barbara oil spill is fascinating reading. It takes a subject, an incident–indeed a tragedy–and examines in minute detail all of its political, ecological, and scientific ramifications. The Steinharts have produced a balanced account of the Santa Barbara disaster without slighting either the environmental issues or interests involved, or those of the petroleum industry, or the governmental agencies which had to sift and decide between many conflicting views. For the student of government, *Blowout* tells a great deal about how government can work positively to solve problems, how it often fails to do so without determined men who are not afraid to buck city hall, and how the Everyday Citizen can have an impact on government to make things happen. Carol and John paint the picture of the disaster in Santa Barbara with such accurate but still vivid colors that the reader cannot help but feel that he was *there* when the whole terrible thing occurred.

Walter J. Hickel

Anchorage, Alaska
August 1971

Preface

"The bad guys wear black hats." We yearn for certainty: for clear-cut right and wrong, innocence and guilt. One of our main articles of faith is that enough research or thought or information will inevitably yield a moral and painless solution to any problem (with help from God, of course, who used to reside in Heaven but now spends most of his time in the laboratory). Both science and long-range plans are necessary, but neither can cope with the unexpected or tell us how to lead our lives. Human aspirations for a meaningful life must determine our actions at each moment.

In this book, we have tried to provide a balanced and comprehensive account of one environmental insult, the Santa Barbara oil spill of 1969. The details are unique to this particular event, but the problems are pervasive and will arise again and again with increasing severity until we deal with them bravely and decisively.

There is no such thing as a complete description of an event. One can only try to distill the essence of what happened. It would have been easy and justifiable to select, from the vast amount of public and private information we reviewed, that material which would support a particular point of view. We chose instead to follow the tradition of science and to present as fairly as possible the most significant facts of the case and to comment on the accuracy with which they are known. We have tried to be objective in our selection and presentation of facts, while allowing ourselves the luxury, now and then, of their subjective interpretation. We are people with feelings and opinions who have studied this oil spill and played a role in the events surrounding it, and we are appalled by the conclusions which force themselves upon us. There are no villains and no blameless victims. The only solution seems to involve such drastic cultural changes that few among us dare to contemplate them. We are on an escalating energy binge, headed for disaster unless we alter our course. The Santa Barbara story is many things to many people. To us, it is a study of modern America and a warning of things to come.

Carol E. Steinhart
John S. Steinhart

Madison, Wisconsin
August 1971

Note to the Reader

In this book we present an analysis of a typically intractable environmental problem which is complexly woven of historical, political, legal, economic, technical, sociological, and ecological threads. Any one of these aspects may be examined further by the reader according to his interests; here we present only a general synthesis of the various types of issues involved.

Chapters I, V, VI, and VII are devoted to a chronological account of events surrounding the Santa Barbara oil spill. After the description of the blowout in Chapter I, the chronology yields to three chapters which present historical, scientific, and technical information necessary for understanding what happened during the months following the blowout and for placing it in historical perspective. Chapters VIII and IX describe the ecological impact of the spill. The first of these chapters discusses the immediate effects on birds and sea mammals, the public outrage and the fury of the press. The second takes a more detailed look at the effects of oil pollution and at scientific studies of the Santa Barbara Channel in the year and a half following the oil spill. Chapter X gives a glimpse of some of the legal and legislative problems involved and the frustrated and frustrating attempts of local citizens to deal with a local crisis. What is involved implicitly in the saga of the oil spill is our contemporary society's "need" for ever-increasing amounts of energy, and Chapter XI highlights some of the sinister implications of our energy addiction. In Chapter XII, Santa Barbara takes its place in the panorama depicting the scope of oil pollution and the outlook for the future.

We have provided two aids for locating the events of the oil spill in time and space: a map of the Santa Barbara area, which is the frontispiece of this book, and the *Chronology of Events* on page xv.

Table of Contents

Chronology of Events

1859	Colonel Edwin Drake drills first oil well in Pennsylvania.
1895	World's first offshore wells at Summerland, California.
1920	First federal lease act.
1921	California's first least act.
1953	Submerged Lands Act passed by Congress.
1953	Outer Continental Shelf Act passed by Congress.
1963	First offshore federal lease sale on Pacific coast.
March 1967	Torrey Canyon oil spill off the coast of England.
November 1967	Santa Barbara County Report issued.
February 6, 1968	Sale of federal leases in the Santa Barbara Channel.
March 1968	Ocean Eagle oil spill off the coast of Puerto Rico.
September 1968	"National Multi-agency Oil and Hazardous Materials Contingency Plan" released.
September 1968	Union Oil Company's Platform A set in Santa Barbara Channel.
January 14, 1969	Drilling begins on well A-21.
January 20, 1969	Inauguration of Richard Nixon.
January 24, 1969	Walter Hickel sworn into office as Secretary of the Interior.
January 28, 1969	Union Oil Company's well A-21 blows out in Santa Barbara.
January 29, 1969	National Contingency Plan activated.
January 30, 1969	GOO organized in Santa Barbara.
February 5, 1969	Oil contaminates beaches of Santa Barbara.
February 7, 1969	Hickel halts drilling and production in Santa Barbara Channel.
February 8, 1969	Well A-21 "plugged."
February 17, 1969	Hickel signs amendment governing liability of oil companies for pollution damage.

February 18, 1969	First meeting of President's Panel on Oil Spills.
February 18, 1969	State of California, City and County of Santa Barbara, and City of Carpinteria submit claim for damages to the Secretary of the Interior.
February 20, 1969	Hickel halts offshore lease sales.
February 22, 1969	Completed wells on Platform A put back into production to reduce rate of leakage.
February 24, 1969	Union Oil Company well A-41 blows out in Santa Barbara.
March 1, 1969	Well A-41 plugged.
March 6, 1969	Oil from Santa Barbara reaches San Diego.
March 21, 1969	New drilling regulations issued.
April 2, 1969	Hickel lifts ban on drilling and production on Santa Barbara Channel leases.
April 10, 1969	Pauly Petroleum and five other oil companies file suit against government regarding channel leases.
May 12, 1969	Meeting of President's Panel on the Future of the Union Oil Company Lease.
June 3, 1969	Report of the Panel on the Future of the Union Oil Company Lease released.
September 16, 1969	Oil spill in Buzzards Bay, Massachusetts.
Fall 1969	First and Second Reports of the President's Panel on Oil Spills released.
December 15, 1969	Pipeline leak at Platform A, about 17,000 gallons of oil spilled.
December 20, 1969	Pipeline leak repaired, production resumed.
November 25, 1970	Secretary Hickel fired.
December 1970	Ecological report released, University of Southern California.
January 1971	San Francisco Bay oil spill.

Blowout

I The Blowout

Every strange effect has some strange cause.
My sons and I have done nothing to deserve this.
It is an inscrutable punishment.
Therefore I weep!

Chuang Tzu

The morning of January 28, 1969, began about like any other winter Tuesday in the town of Santa Barbara, California. Over their breakfast coffee or on their way to work, people were reading or hearing the latest news–which each day, although different, seems monotonously the same. Nine Jews were hanged as spies in Iraq, and Israel vowed revenge. American troops began a final push across the Communist-dominated strip 320 miles north of Saigon. Elsewhere in Vietnam, 42 Communists were killed in skirmishes near An Loc and Rach Gia. Record droughts were turning all Chile into shriveled desert wasteland. In the National Basketball Association, idle Los Angeles held onto first place, a game and a half ahead of Atlanta. Inquiry into the circumstances surrounding the case of the captured spy ship Pueblo continued, while final preparations were being made for the trial of Sirhan B. Sirhan, accused assassin of Robert Kennedy. In Washington, D.C., the new Nixon Administration was gradually settling down to business.

People grow immune to this kind of news. It has little effect on day-to-day life. Perhaps uppermost in the minds of Santa Barbarans that day was the break in the weather. Relief had finally come from the violent tropical storms that, for nine days, had unremittingly ravaged southern California. The worst storms to hit Santa Barbara County for 55 years, the Weather Bureau said, and southern California's worst floods since 1938. The city of Santa Barbara itself had escaped with only moderate inconveniences from the rains, but elsewhere in the county damages were severe. The county's losses were estimated at around eight million dollars, of which two million represented agricultural losses. Hundreds of valuable horses and cattle had been washed away and drowned. Flooding and erosion were so severe around the Santa Ynez River that some predicted the land would not be usable for agriculture for five years. State-wide, deaths from the storm were approaching 100. President Nixon had declared the state a disaster area.

How It All Began

Even in the face of the storm damage that day, probably many Santa Barbarans cast their usual baleful looks at the oil rigs rising ominously from the channel, although many others had managed to blot the monsters from their consciousness in the way Americans have of overlooking ugliness. From Union Oil Company's five-month-old Platform A, five-and-a-half miles out in the Santa Barbara Channel, well number A-21 had been drilled to a depth of 3,479 feet. Drilling was approaching the primary target, a reservoir of petroleum thought to be about 3500 feet below the sea floor. At 9 o'clock the platform team had begun to circulate drilling "mud" in the well, in preparation for the removal of the drill pipe to replace a worn bit. After the drilling was completed, the next step would be to run electrical logs, and then the new well would be completed for production. It was the fifth well to be drilled from Platform A. The Daily Drilling Report for January 28, 1969, indicates that conditions were stable and there was no sign of impending trouble.

The driller began the routine procedure of pulling up the drill pipe in stands of 90 feet, pumping mud into the annulus as each stand was removed. The kelly (the drive mechanism) was disconnected and set aside. As the eighth stand was being disconnected, mud began to flow out of the top of the pipe and splash down onto the rig floor. The time was 10:45 a.m.

When the mud began to flow, three rig hands grabbed the emergency check valve, provided for just this sort of "kick," and attempted to screw it in. Mud roared to a height of 20 feet above the pipe, while below, the men slipped and splashed and wrestled with the 75-pound valve. Then the mud flow stopped as suddenly as it had begun, and a heavy, noxious hydrocarbon mist poured from the open pipe. Within seconds the rig floor enclosure was engulfed in the mist and the rig hands, almost blinded and numbed by fear of fire, abandoned their futile efforts with the valve. Thirty seconds had elapsed. In thirty seconds, you breathe about ten times and your heart beats about thirty-five times. And that was the time from the beginning of the mud flow until the check valve was laid down and the next emergency measure was started. Ten breaths. Thirty seconds.

With visibility almost zero, verbal communication impossible in the roar of gas, and working conditions treacherous in the mud and mist, the driller and rig hands hooked the elevators onto the kelly for stabbing into the drill pipe. As the driller was lifting the kelly out of the rat hole, the hose caught under the drain bull plug and broke it off, thus eliminating the possibility of control through the kelly.

Three more minutes had passed. In three minutes, at a steady pace, you can walk a quarter of a mile.

The engineer instructed the driller to lay down the kelly and go to the last remaining means of well control, closing the hole with the blowout preventer blind rams. The driller lowered the blocks and grabbed the drill pipe to pick up the tension of the pipe column. The hydril blowout preventer was actuated to hold the drill pipe taut while the elevators were disconnected. But the hydril was unable to hold the weight of the pipe. This was not surprising, as the hydril is designed to encircle the pipe for pumping control, not actually to hold the pipe. But would the lip, perhaps, catch on a drill collar long enough for the elevators to be removed? The driller decided to try. He picked up the pipe in the derrick to the next tool joint and actuated the pipe rams, jamming the pipe in the blowout preventer assembly. At last the elevators were removed, the drill pipe dropped into the hole, and the blind rams closed. The well was sealed off at the surface. The drilling time indicator chart on the rig floor showed that thirteen minutes had passed since the start of removing that fateful eighth stand of pipe. Thirteen minutes of business as usual in unsuspecting Santa Barbara. Barely time for a coffee break.

Some ten or fifteen minutes after dropping the pipe and setting the blowout preventer, while the rig crew was proceeding with standard "choke and kill" operations, the drilling engineer observed the first bubbles in the ocean about 200 yards off the northeast corner of the platform. Then large boils of explosive, poisonous natural gas began to roll toward the platform. Workers escaped in boats. After a moment of strange calm, enormous volumes of sepia-colored oil floated up to engulf the area. The Santa Barbara Oil Spill had begun.

Further efforts to control the well from the rig floor would be futile, because flow from the well bore was blowing out of control below the level of the conductor casing, through openings on the ocean floor itself.

As thick oil continued to rise from holes along the fault line on the ocean floor, the oil men may have begun to suspect that they had a very serious problem. Nevertheless, the next step seemed clear. As soon as the platform could be remanned, they would circulate heavy drilling mud in the well at a depth below the blowout, to create enough hydrostatic pressure to prevent further escape of gas and oil. Success was optimistically predicted to come within hours. But problem after problem was encountered, and the hours stretched on into days, until Union Oil officials reluctantly admitted that they did not know when the well would be brought under control.

First, the bottom of the drill pipe was plugged and stuck, preventing circulation of the mud. The obvious solution was to disconnect the offending section of pipe and remove it. But the crew was unable to remove a back pressure valve from the drill pipe. The next step was to mill out the valve. More hours were lost as the necessary equipment was assembled and installed, while Union Oil officials nervously reassured each other and the world that success was certain after the milling was complete. On Monday, February 3–or the sixth day of the runaway oil, depending on how you reckon time—the milling operation was begun and completed. After days of delay, the crucial job took just two and a half hours. The crew began to circulate sea water through the well, but this did not stop the flow of oil and gas. Early on Tuesday, circulation of drilling mud began.

Hopes rose at first as the amount of oil blowing out seemed to be reduced, and for short intervals of time the large blow at the northeast corner of the platform actually subsided completely. About 3000 barrels of mud were pumped into the hole that Tuesday morning, as rapidly as the pump could pump. But after six hours of labor, with rapidly diminishing mud supplies, officials had to admit that it was not fast enough. Not only was the pump too small, but vastly greater quantities of mud would also be needed.

All mud pumping was stopped, while the frantic call went out across California for more—more mud, more mud barges, more holding tanks, and a larger pump. More help. For want of something more effective to do while supplies were stockpiled, the crew pumped sea water into the hole. Around the clock they pumped sea water, 20 barrels per minute. And oil—was it 500 or 5000 barrels per day?—continued to pour from the ocean floor.

In the frenzied, round-the-clock activity—with oil men and newsmen, federal officials and state officials, technical people and just plain angry people all on the scene—conflicting reports and communication failures were inevitable. What *was* the magnitude of the spill? Union officials maintained that the news media were grossly exaggerating, while other observers accused Union of systematically underestimating the damage. *Were* birds and other forms of life being exterminated in enormous numbers? Again, reports varied from essentially no damage to total ecological disaster. Garbled, misleading, and totally wrong information was relayed to Washington, D.C., where confused officials in the brand-new Nixon Administration were trying to determine whether they were confronted by just another brush fire or a full-fledged crisis. For example: on January 29, the day after the trouble began, the Union Oil Company had proposed drilling a relief well as a backup measure in the event that operations on well A-21 itself were unsuccessful. Somehow, in the uncertain process of human communication, the word got relayed to Presidential Advisor Edward Wenk that a drilling ship was already positioned above the well, preparing to intercept the blown well with a new hole in order to seal the well below the fracture zone; and Union officials estimated that the sealing would be *complete within 12 hours.* Actually, the plan was not even approved by the Regional Oil and Gas Supervisor of the U.S. Geological Survey (U.S.G.S.) until the next day, January 30. Two more days passed before the floating drill rig WODECO II arrived from Long Beach, and actual drilling did not begin until Sunday, February 2. The estimated time for completion of the relief well? Ten to fourteen days.

By Friday, February 7, the necessary mud and equipment had been assembled and weather conditions finally allowed work on Platform A to resume. At about 1:00 p.m., the preliminary pumping of "heavy" sea water (sea water weighted with calcium chloride) began, followed by the pumping of mud. At 4:30

p.m., observers on the platform saw some mud rising to the ocean surface in place of oil. At 5:30, the boil near the northeast leg of the platform changed to clear sea water. At 6:30, the boil had diminished to a few small bubbles of gas with occasional traces of oil. At 8:00 p.m., jubilant observers reported no visible evidence of bubbles, and officials decided to complete the job by plugging the entire well bore with cement. By midnight on Friday, 1,159 sacks of cement slurry had been pumped into the hole.

Tests made on Saturday morning showed that the top of the cement column was 960 feet beneath the rig floor, and plans were made at that time to complete the cementing of the well to the level of the ocean bottom. On that day, Saturday, February 8, Union Oil Company's President Fred Hartley released the following to the press: "In our opinion the well has been plugged. As anticipated, minor residual gas is continuing to work its way to the surface. For added insurance the hole will be filled with cement to the surface." The well was sealed; the leak was over. All that remained was cleanup and removal of the oil, a sorting out of what had happened, and plans for measures to prevent its happening again. All that remained was the soul-searching period of convalescence. Or was it?

Oil Comes to the Beaches

Although the round-the-clock efforts of thirty brave men to control the leak on Platform A were the focal point of the action, they were but a part of the total story. Concurrent with work to stop the leak were efforts to contain the oil, to protect harbors and beaches, to rescue threatened animals, and finally to clean up the beaches after attempts to ward off the slick failed. From the beginning, the Union Oil Company had no choice but to accept full responsibility for the cleanup of the spill, although public and private forces were marshaled almost immediately to assist where needed and to coordinate the many-faceted attack on the oil.

Press reports were numerous and dramatic. Witnesses on the scene and vicarious viewers across the nation watched in mingled horror and fascination as the ever-growing slick now advanced, now retreated, according to the whims of wind, tides, and weather. The weather was unusually violent and capricious throughout all of January and February.

The first reports were reassuring: winds were favorable and there was no immediate threat to coastal communities. Besides that, Union Oil began on the morning of January 29 to spray chemical dispersants on the oil. At first this treatment seemed to be effective. But efforts to stop the leak were faltering, and in the early hours of January 30 the wind shifted, driving the oil to within a mile of Carpinteria's glistening beach and threatening to leave heavy deposits between Santa Barbara and Ventura. Miles of log barriers were hastily lashed across the entrances to four endangered harbors and marinas, while a boom was made ready to protect Santa Barbara itself. Meanwhile, on Platform A, hazardous levels of gas and oil necessitated evacuation of all personnel. Conditions were so bad that despite the Coast Guard's policy against use of chemicals, the aerial spraying continued, as the

Federal Water Pollution Control Administration (F.W.P.C.A.) reluctantly granted permission and the Coast Guard acquiesced.

Late in the day, an offshore wind temporarily relieved the threat to the beaches, blowing the gas out to sea. Now the slick extended northeast from the platform in a giant U-shape, its tips menacing the coast. Firemen reported oil in the surf at Rincon Beach, a popular surfing area.

The next day, about 1000 yards of shoreline between Rincon Point and Pitas Point were contaminated, and Union Oil began surface application of dispersants from a 170-foot vessel rigged with booms. But again, favorable winds kept the worst of the slick from coming ashore. Elsewhere, beach patrols reported light oil sinking into the sand without leaving a residue.

The forecast for the night was grim. Once more, operations were halted on Platform A because there was no wind to drive away the gas. By the morning of February 1, oil was beginning to reach the coast in greater quantity and five miles of beaches were contaminated. In the midst of the deteriorating conditions, the F.W.P.C.A. ordered that application of chemical dispersants be stopped, on the grounds that, based on estimates of the amount of oil spilled, chemical use had already greatly exceeded the manufacturer's recommendations. The Coast Guard requested additional pumps, vacuums, surface craft, and boom materials to augment Union's faltering efforts at containment and cleanup. Two reports from Union were discouraging: There was no progress in work on the platform, and delivery of 1000 feet of boom, expected that day, would be delayed. The Coast Guard then recommended that Union begin at once to construct a boom from telephone poles, in order to contain the oil at its source.

The Coast Guard's routine report of February 3 described heavy oil buildup around Anacapa Island, one of the four Channel Islands. Oil had advanced to within 100 yards of the beach at Carpinteria Beach State Park, while light fingers of slick crept past the trash barrier off Pitas Point to within a half mile of shore. Late in the day, Union reported that they had assembled the telephone pole boom around the source of the leak. They were also using logs, straw, and other coagulants in waters off the threatened beaches, to serve as a barrier and to facilitate recovery of the oil. They had positioned additional inflatable booms in coastal areas.

The next day the wind shifted, driving crude oil into the Santa Barbara Harbor. Because of the strong tidal currents typical in the channel, the boom around the platform was not functioning according to expectations; and Coast Guard Cutter Walnut was dispatched to the scene with buoys, anchors, and chains to install a mooring system for it. Meanwhile, the boom which had been standing ready in Santa Barbara was hastily closed, and Santa Barbarans settled back uneasily to await developments.

They did not have long to wait. The next day, February 5, a ship rammed through the barrier. The oil came in. It coated the beaches as far as the high tide line, coated them with between a quarter and a half inch of thick greenish-black goo. It licked onto the hulls of some 750 boats moored in the harbor and forced the evacuation of more than 60 people who customarily slept on their boats. The situation deteriorated rapidly. Still more federal resources were authorized, to spray chemicals on the Santa Barbara Harbor, remove the oil, and reinforce the booms. Beach cleanup crews labored unremittingly, using straw, talc, and "Ekoperl" to sop up the oil.

Adverse weather on February 5 and 6 hampered booming activities and permitted continuous heavy concentrations of oil to reach the shore. The weather also prevented work on the platform. When the weather moderated on the 7th, plans to boom the source were revived. But with the sealing of the well the next day, full attention was shifted to cleaning up the beaches and protecting those not yet contaminated.

Uncle Sam to the Rescue?

The Coast Guard and other federal agencies rapidly became involved in the disaster. At about noon on Tuesday, January 28, Don Craggs of the Union Oil Company notified Lieutenant George H. Brown, III, Commander of Coast Guard Group Santa Barbara, that Union was experiencing gas blowout from a well being drilled from its Platform A. Craggs was confident that emergency actions under way would stop the gas flow without further incident. He therefore notified the Coast Guard only because of the potential fire hazard and rescue involvement, and not from a pollution standpoint. Brown immediately relayed the report to the office of his superior, Rear Admiral Charles Tighe, Commander of the Eleventh Coast Guard District. He then made preparations for the (what seemed to be) unlikely event that his men would be needed for rescue and salvage operations, and went home. For days, the Coast Guard had been busy around the clock rescuing people who had been stranded by the floods of the last week; and the oil incident did not sound particularly alarming.

At about the same time he notified Brown, Craggs reported the trouble to the Los Angeles office of the U.S.G.S., thus triggering the reaction of the sprawling, unwieldy federal machine. Immediately, Harry Cypher, District Engineer, began preparations for a trip to the scene. He arrived the next morning at 7 o'clock. Regional Oil and Gas Supervisor Don Solanas was summoned from a business trip in Wyoming. And in Washington, D.C., control center of the machine, the Geological Survey's Dewey Acuff and Eugene Standley prepared to leave for Santa Barbara along with Kenneth Biglane and Charles Pierce of the F.W.P.C.A.

Even though he was off duty, Brown went over and over in his mind the steps he would take if the blowout turned out to be a significant problem. He would probably have to get an aircraft to Santa Barbara in time to make an air survey before dark. A check with his office yielded the reassuring news that Craggs had not called back. Either the situation was under control or the U.S.G.S. had taken over, as Brown had expected it would. So that evening, relaxing at home, Lieutenant Brown had no premonitions of disaster when he was summoned by a second call from Craggs. The bubbling gas was now mixed with oil, and a pollution situation was rapidly developing. Again, Brown notified his superior, this time requesting an aircraft for aerial surveillance beginning at dawn on January 29. He did not get much sleep that night.

The survey was completed at 9:00 a.m., after which Brown was briefed by Union Oil Company officials. The results of the survey and the briefing came out at 11:20 a.m. in the Coast Guard's "Oil Spill Santa Barbara Sitrep One":

> Union Oil Platform A in position 34-19-55 N 119-36-49 W experienced high pressure gas blowout delivering gas and mud from the end of drill pipe. . . . Emergency action taken. . . . Gas pressure now being relieved through natural faults in earth crust. . . . Union Oil attempting to retrieve drilling pipe and shut in well. . . . All agencies notified. . . .

It was the first of many such Situation Reports, no one could suspect how many, to be issued during the course of many months.

The course of federal activities was largely as prescribed by the "National Multiagency Oil and Hazardous Materials Contingency Plan" of September 1968. This was the first major test of the plan, which had been conceived and drawn up during the Johnson Administration. It was developed at the request of the President himself, who, like many presidents, urgently felt the need for coordination and cooperation among the plethora of relatively autonomous federal agencies, especially in times of crisis. The disaster following the grounding of the tanker Torrey Canyon in March 1967, with the subsequent mismanagement and confusion, can be said to have fathered the Plan; but its gestation period was certainly shortened when, two weeks short of one year later on March 3, 1968, the Ocean Eagle foundered off San Juan, Puerto Rico. Perhaps there was even less time than people had thought. . . . In the words of the Plan itself:

> The development of a national awareness and concern over the hazards and damages to water related resources can be traced in large part to the sinking of the tanker, Torrey Canyon. This marine casualty off the south coast of England caused a massive oil spill with a tragic destruction of water related resources along both the English and French coasts. But more importantly, it galvanized the United States into action and caused us to assess our own capabilities to cope with massive spills of oil or other hazardous substances. President Johnson recognized this need by directing the Secretary of the Interior and the Secretary of Transportation to study the nation's capabilities for handling such disasters. . . . (Their) report documented our unpreparedness to deal with a catastrophic oil spill, and clearly pointed to many areas of needed Federal action.

The President then

> . . . directed the Secretaries of Defense, Interior, and Transportation and the Director of the Office of Science and Technology to assume special responsibilities in order to strengthen this Nation's preparedness to act in the event of an oil spill pollution emergency along our coasts and waterways. The President further directed the Secretary of the Interior to assume primary

> responsibility for completing . . . multiagency contingency plans for responding to oil spill emergencies.

The national plan called for regional plans and set guidelines for their establishment. In Santa Barbara, however, regional operation was still under California's Cooperative Agreement for Marine Chemical Spill Disasters, which antedated the National Contingency Plan and did not conform to its specifications. Now, six months after the birth of the national plan, we can see both it and California's obsolete regional plan in action in Santa Barbara.

After issuing his first situation report, Lieutenant Brown began to apprise local government officials of the situation. A second air survey, completed at 1 o'clock, corroborated the hasty observations of the first. The slick extended 12 miles south and southeast of the platform. After the results of this survey were announced, about 5:00 p.m. on January 29, the machinery was put into motion to implement the regional plan, establishing on-scene headquarters at the Coast Guard Group Office, Santa Barbara. In Washington, D.C., the Joint Operations Team (J.O.T.) was alerted, activating the National Contingency Plan. Lieutenant Brown was designated On-Scene-Commander (O.S.C.), and his ordeal had officially begun.

At 9:00 p.m., Brown held a briefing of local, state, and federal officials and representatives of the Union Oil Company. The government officials approved Union's strategy for dealing with the crisis and established procedures to ensure rapid dissemination of information on the progress of the slick. By then an estimated 150,000 gallons of oil had escaped, and the Eleventh Coast Guard District had arranged to augment the Santa Barbara forces for a possible extended operation.

At 1:30 on the afternoon of January 30, the California Regional Operations Center (R.O.C.) was activated in San Francisco. (The Santa Barbara incident highlighted a number of flaws in California's out-dated regional plan, not the least of which was the location of R.O.C. so far from the disaster area and in the 12th, not the 11th, Coast Guard District. Later on, in fact, the R.O.C. would be transferred to Los Angeles.) At their first meeting, R.O.C. members reviewed actions taken and proposed so far, pronouncing them correct but inadequate. They suggested that Lieutenant Brown and Union Oil use floating equipment and pumps to remove surface oil near its source. They also recommended that additional federal resources be made available to Union Oil on a reimbursable basis.

In Washington, Interior Secretary Walter Hickel began a review of federal drilling regulations and announced to the press that his team of Acuff, Standley, Biglane, and Pierce was California-bound to assist wherever possible. At the scene of the action, Coast Guard Cutter Pt. Judith was policing the precautionary zone within a five-mile radius around Platform A, air surveillance flights continued, and beach patrols were established. Now in its third day, the oil spill was officially classified a "pollution incident and a potential major disaster." Acuff, Standley, Biglane, and Pierce arrived on the scene in the still-dark morning hours of Friday the 31st, to be briefed by Solanas, Cypher, and Union Oil officials. By then, a veritable army of public and private forces had been marshaled to deal with the disaster. But deal with it they could not, for what could they actually do?

They made surveys, they issued reports, they established intricate and unreliable communication networks, they held meetings, and they hassled over policy and procedural matters. All this had very little effect on the recalcitrant oil. For the most part, all the concerned spectator-participants could do was to look on helplessly while the Union Oil Company summoned the skill and know-how of the industry to battle. Union accepted full responsibility for cleanup and control, although were it not for the laws and the relentless insistence of public officials, they might have been content merely to extricate themselves from their embarrassing predicament far enough to escape further public notice. But no matter whether out of necessity or magnanimity, they did rush oil-drilling expert Coot Matthews to the scene; they hired a crew from Houston's Red Adair Corporation, famed for fire fighting and capping gushers; and they committed themselves to boom off all harbors and sloughs as necessary and to contract for cleanup. That their efforts were inadequate may have been more an observation than a condemnation.

It may be well to recall that the Nixon Administration officially began on January 20, just eight days before the blowout. Secretary of the Interior Hickel had taken office only four days before, after undergoing a most grueling investigation at the hands of a hostile Senate. Of all the Nixon Cabinet appointees, it was Hickel who aroused the most bitter feelings and anxieties among the more liberal elements of the Senate and the general populace. He was particularly suspect for his cavalier attitude toward the environment; after all, he had flippantly condemned "conservation for conservation's sake." Hickel was proud of his toughness, the traditional toughness of the American frontiersman. In the severe environment of frontier Alaska, his shrewdness and perseverance had brought him from poverty (the "I arrived in Alaska with 23¢ in my pocket" kind of thing) to wealth, power, and finally, to the governorship. But now, senators feared that he was so business-oriented as to be insensitive to conservation of the nation's resources, of which he had been nominated as chief custodian. Furthermore, there were serious questions concerning his oil lease and stock holdings in Alaska. It was all summed up by South Dakota's Senator George McGovern, who had dreams of some day relieving Nixon of his job: "It is my present judgment that Gov. Hickel does not have the background, the philosophy, or the insight to meet this responsibility so essential to our nation's future." Thus it was that the appointments of the other eleven Secretaries were unanimously and routinely confirmed at the Senate's first meeting following the inauguration, while Hickel's confirmation remained uncertain for several days. To mollify the senators and push through his man, President Nixon allowed the news to leak that he was appointing Russell Train, noted conservationist, as Undersecretary of the Interior. Hickel was finally confirmed unenthusiastically by a vote of 73 to 16, and was sworn into office on January 24.

Throughout the new government at the hierarchical level just beneath the department Secretaries, officials had either just reported to their new jobs or had not even been appointed yet. It was part of Nixon's policy to take it easy, play it by ear, let his new Secretaries select their own co-workers. Those appointees already on board were still unfamiliar with their new departments and with each other. Perhaps worst of all, they were confronted by the formidable array of top-level career civil servants who are sometimes more interested in protecting themselves than in helping to solve the problem at hand. The game of finding out whom to trust was going on in earnest.

Like most administrations on the way out, the Johnson Administration had not come suddenly to a stop on its last day. Over the course of weeks and months it had slowly run down, as top-level officials left, one by one, to resume their old careers or to take up new ones. Those who were left bided their time to see what changes in policy the new administration would bring. Nixon did not begin with a storm of activity. He had announced his goals clearly enough: law and order, moderation, private enterprise, peace, a strong national defense. But how he planned to move toward these goals was still a mystery. The predictions for a slow-starting administration were being fulfilled. So in Washington, the problem of the oil spill was first of all a problem of how to gear up suddenly for an unexpected emergency, with people new both to the problem and to their jobs. It has been said that if you want to start a revolution, do it in the first few weeks of a new administration. Well Number A-21 could hardly have picked a more propitious time for its revolt.

Added to the general confusion in Washington was the confusion of the reports that arrived throughout the whole first week of the spill. There was a blowout and oil was pouring from the well. There was a blowout and oil was pouring from the sea. There was no blowout at all. One well blew—no, several. Oil was leaking around the platform, but the leak would be shut off within six hours. The weather was bad, and six hours more would be needed. And six hours more and six hours more. In each succeeding report, the leak would be stopped in just six—or perhaps twelve—hours more. Clearly, most of the information being received was wrong. But somehow, Washington was expected to respond appropriately to the crisis. That Washington, and specifically the Department of the Interior, did respond promptly and decisively suggests that those critics of Hickel who condemned the frontiersman aspects of his character were correct in one respect. Hickel exhibited the most important ingredient of the frontiersman: courage.

Another newcomer in Washington who was to influence the course of events in Santa Barbara was Lee DuBridge, Science Advisor to the President. Unlike Hickel, DuBridge entered into his new office without controversy—in fact, almost without notice. His appointment did elicit nods of satisfaction from the Scientific Establishment, for DuBridge was the Establishment Personified. He had long been a member of the inner circle of the trusted scientific elite, first as director of M.I.T.'s Radiation Laboratory, wartime contributor to radar research and development, and later as president of the California Institute of Technology.

As Science Advisor, DuBridge was also Director of the Office of Science and Technology (O.S.T.), some of whose staff hoped to have a say in future offshore drilling policies. DuBridge, as spokesman for O.S.T., was therefore most eager to assume a key advisory role to the President and the Secretary of the Interior in the Santa Barbara incident. Furthermore, Secretary Hickel had not yet selected his science advisor, and it is probable that DuBridge was eager to influence this choice. It was with these hopes and ambitions in his mind, then, that DuBridge suggested to the new Secretary that O.S.T. undertake to convene a panel of nongovernment experts on petroleum geology and related matters, who would advise on the most promising ways to prevent future Santa Barbara-type disasters.

One old timer among the newcomers in Washington was William Pecora, Director of the U.S. Geological Survey. Pecora had 30 years of distinguished service with the U.S.G.S. to his credit, and was held in high esteem by his colleagues in science and government. For the recriminations associated with Santa Barbara, Pecora was the most senior scapegoat. Secretary Udall was no longer around to take the rap; and Secretary Hickel, although many people expected him to take the wrong action, could scarcely be blamed for the original blowout. But Pecora—Pecora was in command of the agency that was most intimately connected to the original leasing in the Santa Barbara Channel and the subsequent management of operations there.

When appointed as director in 1965, he had taken over a staid and scientifically respectable agency with a history of conservatism. He succeeded several decades of directors distinguished mostly by their inactivity. Pecora, a former Olympic athlete, liked action and the personal engagement of Washington politics. His activity in policy matters and in defense of U.S.G.S. programs brought his agency into the rough everyday political world. Critics point out that Pecora was a scientist, not a seasoned administrator. True, his biography does lean to degrees, scientific societies, research honors, and membership in national academies, but, good poker player that he is, he took bad deals well; and his appointment in 1971 as Undersecretary of the Interior was helped more than hurt by his performance in Santa Barbara.

On February 2, Secretary Hickel himself flew to Santa Barbara to view the extent of the damage and decide what actions his department might take. By then, the groups participating in the cleanup and control efforts included the Weather Bureau, the Coast Guard, the F.W.P.C.A., the Geological Survey, the Army Corps of Engineers, the Public Health Service, the Audubon Society, the Sierra Club, the local Humane Society, the California Department of Fish and Game, the U.S. Forest Service, the University of California at Santa Barbara, and, of course, the Union Oil Company. No wonder that Hickel, who was doomed to bear the brunt of the criticism for whatever happened in the future, decided to see for himself what was going on. He was unpopular in Santa Barbara before he even got there, before he did anything at all. "They've sent the fox to inspect the chicken coop," was how the newspapers worded it, and—"Nixon is President, but Oil is King."

Hickel arrived in Santa Barbara on the evening of the 2nd, and finding the situation even worse than he had expected, he ordered an immediate review of the drilling and casing procedures being used for drilling operations in the channel. The official board of inquiry, composed of Standley, Acuff, and Solanas, began its work the next day. Meanwhile, Hickel, in a bold move, asked all oil companies that were actively drilling under federal leases in the channel to suspend their operations until the review was completed, and they were given clearance to resume. Six major oil companies were involved in this voluntary action: Union, Humble, Phillips, Gulf, Texaco, and Mobil. Only Union's relief well was excepted from the temporary drilling suspension. Hickel's action came as a surprise, because, although he had been bombarded with demands to stop drilling in the channel, his remarks on his arrival had indicated that he had no such intentions. "DRILLING HALTED" was the simple but eloquent message that leaped from the main headline in February 4's *Los Angeles Times.*

The government review of drilling practices was carried through quickly. It recommended additional requirements for blowout prevention equipment testing, more stringent mud control during all operational phases, confirmation of storm choke and well safety devices in the completion programs, additional casing requirements, and stricter consideration of casing variance justifications. Despite the recognized need for stricter regulations, however, by February 5 all the companies had been cleared by the Interior Department and had resumed drilling. The shocked Californians howled "Judas!"

If the oil companies understood their clearance to be an all clear signal for future operations, they were mistaken. For only two days later, on February 7, Secretary Hickel ordered a halt to all drilling and production in the channel. The order went out in telegrams to the heads of the five companies which were then operating offshore drilling rigs and producing wells on federal lands off the coast of Santa Barbara: Union, Humble, Texaco, Phillips, and Mobil. Again, only Union's relief well was excepted. Announcing the order, Hickel said, "I have directed all of the oil companies involved to reduce their platform operations to a standby condition until it can be shown with some certainty that we are not risking another blowout in the area by continued drilling. Our Department engineers produced preliminary data which indicates [sic] that drilling alone may not pose a threat of a blowout. However, it has become increasingly clear that there is a lack of sufficient knowledge of this particular geological area. This lack leaves us no other reasonable course of action than to halt drilling until Union's ruptured well can be sealed, and until the required geological knowledge is secured."

News of the drilling halt was greeted by a jubilant press and citizenry whose relief was not unmixed with an "I told you so" attitude. It was not an easy decision for Hickel to make, even though public sentiment supported him and existing laws and Interior Department regulations granted him full authority to issue such an order in the event of a pollution incident. Maintaining the oil rigs even in standby status is a costly business. John Fraser, Vice-President of the Union Oil Company, estimated that each day of the drilling halt cost the oil companies $300,000. Although Fraser's figure may contain some questionable factors like the value of the oil the companies did not produce that day, it is of the same order of magnitude as other estimates. All oil companies drilling in the channel had to absorb these losses; their only recompense was an extension of their leases, in some cases by as much as 120 days. Hickel's move in halting the drilling showed perhaps more courage than wisdom, considering his uncertainty about what to do next.

In accordance with the pervasive human tendency to "When in doubt, set up a committee," Hickel, as he announced the drilling halt, also established not just one committee, but two. One group, acting on the recommendations of the Solanas, Acuff, and Standley report, set out to revise, update, and toughen the drilling regulations. The second group, which was a joint venture of the Justice and Interior Departments, was merely to investigate the situation and make further recommendations.

Yet a third group was to be established. Acting on Science Advisor DuBridge's offer of assistance, on February 11, Hickel suggested to the President that he ask DuBridge to assemble a panel of nongovernment scientists and engineers. The charge to the panel, whose formation was announced two days later by DuBridge, was fourfold: to consider (1) how the federal government could best and most rapidly assist in restoring the beaches and waters around Santa Barbara; (2) how best to prevent future incidents of sudden and massive oil pollution; (3) how existing technology could be developed to control oil pollution; and (4) in what avenues of research could federal resources be most profitably committed toward solving the oil pollution problem.

The formation of the panel had barely been announced when news of a new oil leak in Santa Barbara went out across the UPI wires to a pollution-weary nation:

> Oil is pouring into the sea again from a fissure in the sea bottom near an offshore well in Santa Barbara Channel, federal authorities said yesterday. The flow is a fraction of an earlier leak, officials said. But by nightfall the new leak had formed a streak of oil eight miles long. It began, as did the first leak, under Union Oil Co. of California's platform five and a half miles seaward from Santa Barbara Harbor. It stretched two miles toward the shore, 100 feet wide and an eighth of an inch thick, and then curved northwest parallel to the shoreline as a thin, black streak.

Union Oil Company officials admitted that they had known about the new leak for two days before its presence was confirmed by the Coast Guard. Lieutenant Brown had been forced to reduce the number of his surveillance flights because of lack of staff and aircraft. Unfortunately, this cutback came at exactly the wrong time.

No sooner had a Coast Guard surveillance flight confirmed reports of the new leak than Brown requested Union Oil to send its Sea Sweep barge to remove the oil from the water at the source of the leak. The Sea Sweep was a booming and skimming device rigged up for the occasion. It consisted of 800-foot sections of steel pipe 20 inches in diameter, forming a large V. Each arm of the V was towed by a tug through the oil. At the vortex of the V, a large barge sucked up the scum and transferred it to holding tanks on another barge.

It was a nice idea, but it didn't work. The next day, February 15, the Sea Sweep returned to Santa Barbara with mechanical difficulties. On February 17, the

Union Oil Company announced that their Sea Sweep was ineffective and was being dismantled. Meanwhile, a submarine and divers that had been sent to survey the ocean bottom in the vicinity of the platform reported that they were unable to assess the situation because of zero visibility on the ocean floor, due probably to the material washed into the ocean during the heavy storms.

And the leak continued unabated.

II Oil and Santa Barbara, Partners in History

Of all legacies of the ancient seas the most valuable is petroleum. Exactly what geologic processes have created the precious pools of liquid deep within the earth no one knows with enough certainty to describe the whole sequence of events. But this much seems to be true: Petroleum is a result of fundamental earth processes that have been operating ever since an abundant and varied life was developed in the sea—at least since the beginning of Paleozoic time, probably longer. Exceptional and catastrophic occurrences may now and then aid its formation but they are not essential; the mechanism that regularly generates petroleum consists of the normal processes of earth and sea—the living and dying of creatures, the deposit of sediments, the advance and retreat of the seas over the continents, the upward and downward foldings of the earth's crust.

Rachel Carson

In the catalog of processes by which the surface of the earth is endlessly reshaped, the events at the margins of the Pacific Ocean basin are reckoned among the most rapid. Near the shores of the Pacific—in the South Pacific, in Japan, in the Aleutian and Kurile Islands, and along the coast of California—are the continuous and interwoven bands of tectonic faults. These breaks in the earth's crust have associated with them most of the world's major earthquakes. Yet even the largest earthquakes are tiny adjustments, as footnotes and afterthoughts, to the strains of slow movement and deformation of the rocks. Beyond the fault zones and nearer the land masses around the Pacific is the ring of fire. Here in the ring of fire are found the largest number of the world's volcanoes. Where no wisp of smoke or pool of molten rock can now be found, the mute records in sea cliffs and mountain sides

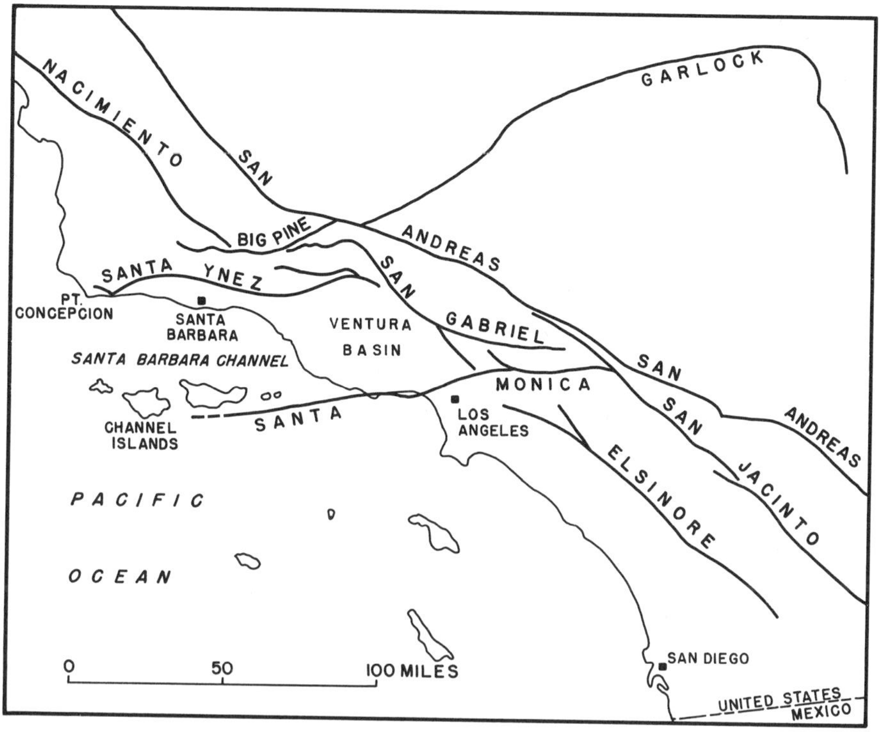

Figure 1. Map of the major faults of Southern California. Based on the index map, figure 1, in U.S.G.S. Professional Paper 679 (1969).

show the present calm to be only the interlude between eruptions of fire and flowing lava.

On the south coast of California one of these great Pacific margin fault zones, the San Andreas, is exposed on land, instead of (as is usually the case) being hidden at the silent sea floor beneath a blanket of oceanic ooze. From Cape Mendocino until they are lost beyond the desert of Southern California's Imperial Valley, the interwoven traces of the San Andreas fault zone are better known than almost any others on the Pacific margin. Southern California's web of major faults is represented in figure 1. The San Jacinto fault, San Gabriel, Hayward, and Elsinore faults, Nacimiento, Big Pine, and Garlock faults, White Wolf fault, and the San Andreas itself are names that call up to the seismologist the shuddering of earthquakes past and yet to come. To the geologist, they are the edges where one giant slice of the earth slips northward relative to another. Santa Barbara, and all the westernmost coast of southern California with it, is moving northward at the rate of one or two inches per year relative to the rest of the United States.

Geologic History of the Santa Barbara Channel

Along the San Andreas fault zone, not all the faults parallel either the coast or the main San Andreas break itself. Some branch at angles to the main fault trend and present a fracture pattern like the crazing of a crack in a piece of antique china. From the San Andreas and San Gabriel faults north of Los Angeles, the Santa Monica fault trends almost due west into the ocean, passing near (or perhaps through) the Santa Barbara Channel Islands. About forty miles further north, the Santa Ynez fault also trends west from the San Gabriel fault to the sea, skirting the southern edge of the Santa Ynez Mountains. In this area, bounded by the two faults, lie Ventura, Santa Barbara, and the Santa Barbara Channel.

This is the Ventura basin. Rapidly, even by the scale of Pacific margin movements, the rocks of this area have been deformed and uplifted several thousand feet in the last few million years. Were it not for the rapid rate of erosion and the relentless action of the sea to distribute the sedimentary debris, this area would be an east-west trending mountain range. The oldest rocks found here date from the Cretaceous Age of Dinosaurs, about 150 million years ago. This is only the most recent one-thirtieth of the earth's history. Earlier rocks are absent from the record. They may have been remelted, or deformed beyond recognition, or buried beneath the deepest erosion valleys beyond the reach of the deepest drills. These earlier rocks may be seen exposed in the Santa Ynez and San Rafael Mountains, where enormously thick sections of later rocks have been eroded away. They are the dark twisted beds laid down on the long steep slopes, far offshore, where the continent's edge fades away in a slope to the abyssal floor of the ocean. These rocks have been moved toward the land for long distances, and the forces that moved them are recorded in their contortions.

Periodically since these early times, the Santa Barbara Channel area has been uplifted and contorted, then eroded flat, only to be invaded once more by the sea to begin anew deposition of sediments. Sometimes the deposits were the sandstones and shales of shallow and intermediate depths. Sometimes the subsidence and deformation were so severe that limestone and phosphate rocks were deposited in water depths of 4000 feet or more. During the time since the mid-Cretaceous, more than 12 miles of rock thickness have been deposited.

It was recently, as such things go, that the particular events of folding, faulting, and erosion occurred to build the Rincon fold sequence from which the Santa Barbara oil spill came. About three million years ago, when our evolutionary progenitors were already noticeably human, a linear trend of folds and domes (the Rincon trend) was slowly squeezed up in the area that is now the Santa Barbara Channel. It is difficult, based on human scales and experience, to imagine the magnitude of the forces and the weakness of the obdurate rocks involved. If the Santa Barbara Channel rocks were scaled down to the size of a card table, they would be as weak and as easily deformed as a thin, watery mud. Uplift movements of the rocks, coupled with subsidence elsewhere, have been so extensive that had there been no erosion, the relief within the Santa Barbara Channel alone would be twice that of Mount Everest.

And the sea continued first to advance and then to retreat. There were once land bridges to the Channel Islands over which the Pliocene elephants wandered freely from the islands to the mainland. Were they stranded when the seas came back? Perhaps, but it is a meaningless question now. They are all extinct, and from their bones in asphalt pits men try to visualize a world that is not part of their cultural memory. Today, the islands exhibit differences in flora and fauna indicative of their eons of isolation from the mainland. In our blundering greed and stupidity, we have begun to obliterate the indigenous biology of the islands, but our success so far has been limited. Fortunately, there is still time to preserve much of what is left.

Petroleum Geology in the Channel: Usual and Unusual

About two million years ago, in the middle of the Pleistocene epoch, the sedimentary rock floor of the Santa Barbara Channel began to deform and a series of wrinkle-like folds appeared. These east-west trending lines of folds and domes are the sites where there might be accumulations of oil. Oil, being lighter than water, works its way upward wherever it can. Along small cracks or fault line rubble, or most often through the permeable sands and sedimentary rocks, oil migrates upward until it either is sealed off by an impermeable layer of rock or escapes into the ocean. Alternate layers of permeable and impermeable rock may be stacked like a sandwich, and at the highest points of any of these folds there may be oil. Thus, multiple pools of oil often occur in a vertical structure, each pool lying beneath its own obdurate and impermeable ceiling of rock.

We already know that some of these fold peaks contain commercially producible oil. One fold trend has produced oil continuously since the first well in the Ventura basin east of Ojai, in 1866. The first offshore field, Summerland, lies on the same fold trend. Further south near Ventura are the San Miguelito, Rincon, and Ventura oil fields, the most productive fields in the area (yielding over 80 percent of all Ventura basin oil up to 1970). This is the shoreward end of the Rincon fold sequence. Figure 2 shows the locations of producing oil fields on the Rincon fold trend, as of 1970. Just offshore is the Carpinteria offshore field. This field, opened in 1966, now rivals production from the Ventura field. To the west of the Carpinteria offshore field is another wrinkle along the sequence of folds. For this elongate dome, the layers known to seal in oil pools come even nearer the surface. This suspected oil field was named the Dos Cuadras oil field (in Spanish, "Two Blocks"). It was so named because the promising portions lay beneath two adjacent block-shaped oil lease tracts, OCS P-0240 and OCS P-0241. Beginning only a few feet beneath the ocean floor, a sequence of oil-bearing sands arches up in an elongated elliptical dome. It could be the best prospect on the entire Rincon trend.

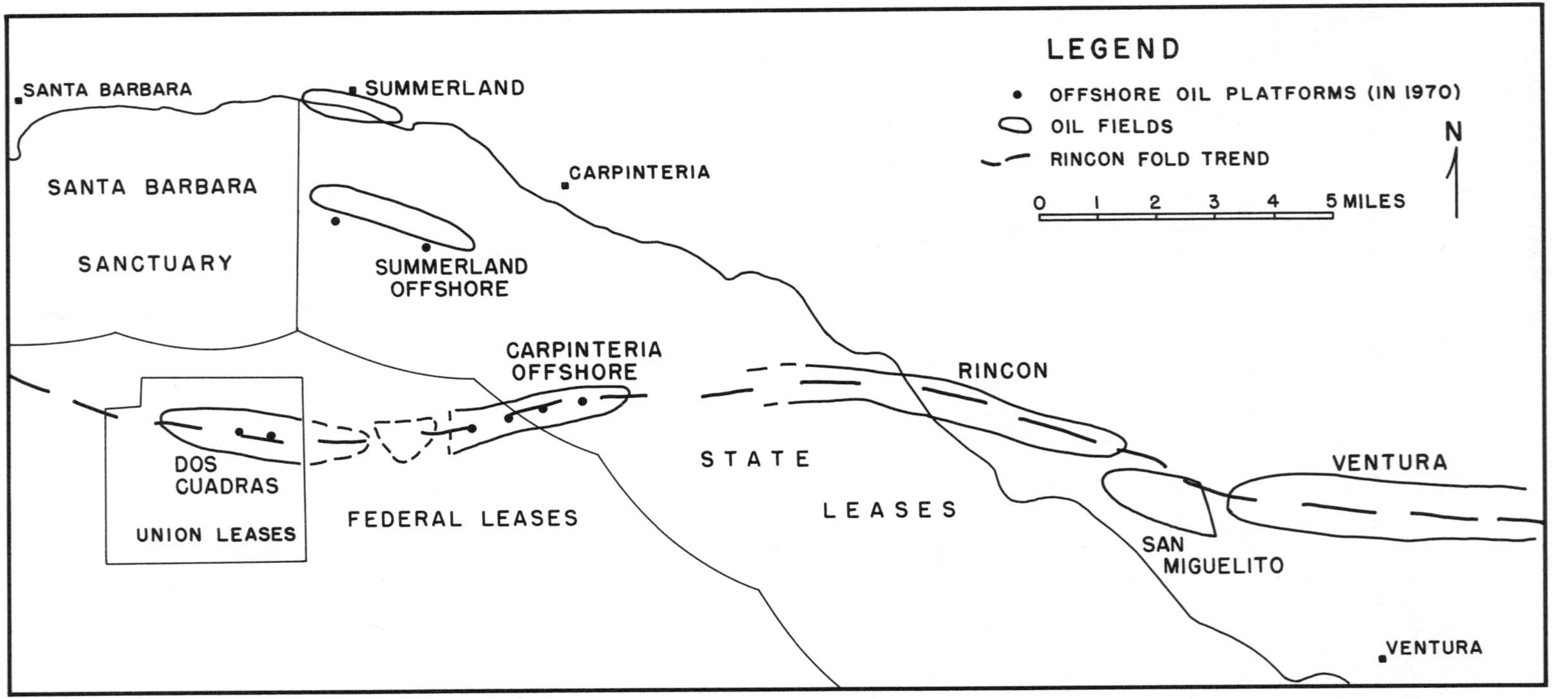

Figure 2. Producing oil fields in the Santa Barbara area. The offshore wells at Summerland were the first to be drilled anywhere in the world. (Based on U.S.G.S. Professional Paper 679 [1969].)

It is likely that the Union Oil Company thought so, when it and its partners bid a record $61.4 million for tract P-0241, in which the next year, the Santa Barbara blowout would occur.

To the east of the trend, the most recent sediments are much thicker than at the Dos Cuadras field and they contain little or no oil. Erosion has been active over the Dos Cuadras field, however, so that less than 300 feet of the most recent sediments remain as caprock to seal in the oil. Then, at about 300 feet, the top of the "Repetto formation" is encountered. This is one of the porous formations along which natural oil seeps have been identified for years.

The caprock above the Repetto formation is almost 90 percent claystone. This rock is so soft and poorly cemented that some samples can be crushed by hand. Nevertheless, it is impermeable to oil, and the oil sands below are rich in oil. Beneath this level, the rock sequence consists of a series of sands and sandstones bearing oil, alternating with thin but numerous impermeable layers of siltstone and shale. Because thick, hard beds are rare (the thickest are but 50 feet thick), cementing the casings of oil wells requires great care. Oil production is possible in more than 60 percent of the interval from the ocean bottom to a depth of 4000 feet. It was this field that was being put into production when well A-21 blew out.

Beneath the present operating depths there is probably still more oil. Comparison with the fields to the east (Ventura, Rincon, and others) suggests that older and deeper strata of the Miocene epoch will also contain oil. This would further enhance the value of the Dos Cuadras field.

The cross section of the Dos Cuadras oil field (figure 3) gives an idea of the sequence of rocks and the extent of oil accumulation. The pressures and readjustments that thrust up these folds at times simply shear the rocks. All the Rincon trend folds have faults in them. Two faults are shown in the cross section. These offsets complicate the position of oil pools and at times the offset of two impermeable beds forms another kind of structural oil accumulation.

During the Santa Barbara oil spill, much public controversy arose over the existence and significance of faults found under Platform A. The presence of such faults is no surprise. Major oil fields often contain faults both large and small. Some fields produce oil primarily from the oil traps formed by faults. In any case, the large faults seem to pose no major or unusual danger for drilling or for leakage. That a pathway along one of the fault zones was forced open by the events of the blowout seems likely but cannot be proved. It matters little, in any case, whether an old fault zone was forced open, because the pressures generated at the time of the blowout were sufficient to crack open the caprock at a new place if there had been no preformed avenue of escape.

Thane McCulloh, of the U.S.G.S., has presented a great deal of detailed information about the geology and physical characteristics of the shallower sections of the Dos Cuadras field in *Geological Survey Professional Paper 679*. The caprock is indeed friable and weak, but before the blowout it did effectively seal in the extensive oil reservoir just below. Only minor natural seepage, much less than the natural seepage at Coal Oil Point, occurred before drilling began. Had an open

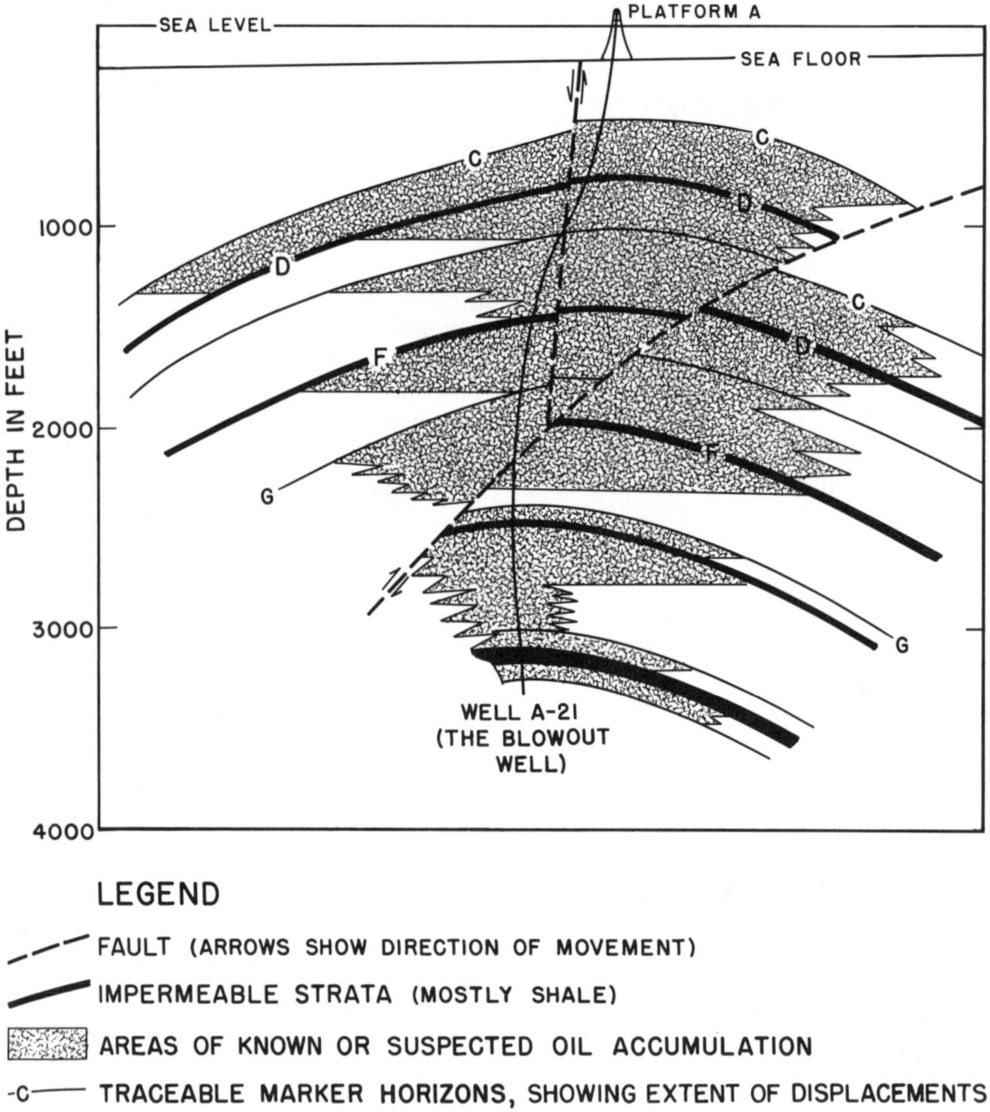

Figure 3. Cross section of the Dos Cuadras oil field. (Based on U.S.G.S. Professional Paper 679 [1969].)

fracture or fracture system existed before the blowout, the oil just beneath would have long since escaped.

Permeability is a measure of the ease of migration of fluids through porous rocks. Although individual pieces of the caprock may be quite permeable, the overall effective permeability is low. The caprock contained small amounts of oil, but not enough to produce commercially. This oil was discovered in the caprock at the time test holes were drilled for the foundations of Platform A. Further showings of oil and gas when the pilings were driven for the platform, as well as loss of mud circulation when drilling through the caprock, indicated that the caprock either was easily fracturable or contained some segments with high permeability, or both. In hindsight, it appears that the caprock was difficult to deal with. In hindsight, it may have been wiser to leave that particular field alone, or to follow a different development plan. In hindsight.

The trouble with the caprock and its deficiencies is not necessarily ended when drilling is completed. As the oil is withdrawn there is the possibility that the crest of the structure will subside, even with a careful program to inject water at the edges of the field to maintain the pressure. The oil field at Wilmington, California, which is similar in some ways to the Dos Cuadras field, has subsided almost 30 feet. Such a large subsidence at Dos Cuadras could fracture the caprock still more, enhancing the chances of new leakage of residual oil. The risk is difficult to evaluate, and the experts disagree about it. McCulloh considers subsidence a definite risk; other experts do not. One thing seems certain. If there is oil to be found in deeper reservoirs than those that have been drilled to date, exploiting these deeper reservoirs will increase both the risk and the potential amount of subsidence. It will not be easy to forfeit potentially large and profitable yields from the Miocene strata below. But would the oil so obtained be worth the risk? Little has been said about these suspected deeper pools of oil. Let us hope that a great deal will be said before deeper drilling is attempted on this structure, and that a sober evaluation of the risks will be made public.

Santa Barbara, Past and Present

Today Santa Barbara, county seat of Santa Barbara County, is a wealthy and beautiful city of about 70,000. She perches proudly on the coastal shelf between the Pacific Ocean and the Santa Ynez Mountains, 100 miles northwest of Los Angeles. People of social position and leisure have flocked to her to build great landscaped estates with palatial mansions–veritable modern castles. The "common folk" of Santa Barbara grow prosperous from the tourist industry or the flourishing agricultural and floricultural trade. One of Santa Barbara's proudest commercial enterprises is the export of flower seeds–as much as three quarters of the nation's supply. Her catalog of tourist delights includes garden tours, aquaplaning, archery and other games and sports, boating and yachting, fishing, hunting, horseback riding, swimming, surfing, water skiing, parks, camping and picnic areas, and an almost constant succession of fairs and festivals. But increasingly during recent years it is oil that has come to dominate the scene. Dominate it to be sure by bringing dollars (100 million of them in 1967), but, concomitantly, by bringing grief, anger, and frustration.

Santa Barbara is an ancient human dwelling place, antedating her name by several thousands of years. Oil, tar, and gas have been natural features of the channel region and companions of its inhabitants for just as long. Probably the first people to inhabit the area, at a time preceding the Bronze Age in Europe, were the "Oak Grove Men," so named because they lived in the humid, oak-forested crests of the Santa Ynez foothills.[1] They were Mongoloid in type, heavy-set, slant-eyed, with flattened noses; and they may represent the earliest culture to exist in what is now the mainland of the United States.

About 2000 years ago, a new race came to the channel area, bringing a new culture. These "Ancient Ones," as their descendants came to call them, were a hunting people who used asphalt from natural seeps to make fast the points on their weapons. The Ancient Ones, the hunters, had the channel region to themselves for about a thousand years, during which time they gradually gave up their nomadic habits in favor of a more settled mode of life. Some time during the 10th century, A.D., another race apparently arrived, eventually intermarrying with the Ancient Ones and merging the two cultures. The descendants of these merged peoples were later named "Canaliño" (channel) Indians by the Spanish explorers, and they became the neophytes of the Santa Barbara mission.

It was 50 years to the month after Columbus had discovered the New World, when two tiny caravels, La Victoria and San Salvador, sailed up the California coast from Mexico and into the Santa Barbara Channel. Juan Rodriguez Cabrillo, a Portuguese navigator sailing for the Spanish Crown, was leading the first band of European explorers along the California coast. A fleet of large Indian canoes, made of boards lashed together and caulked with asphalt from local tar pits, swarmed out to greet the visitors. Their ingeniously constructed canoes were the pride of the Canaliño men.

European visitors to the channel were infrequent for the next two centuries. In 1602, Sebastian Vizcaino set sail from Acapulco to explore the California coast for potential ports. With him was a Carmelite friar, Father Antonio de la Ascencion, who made it his practice to name each new place for the saint on whose festival day the discovery occurred. Therefore, when they entered the channel first charted by Cabrillo 60 years before, the channel and adjacent shore were named in honor of Barbara of Nicomedia, Third Century Virgin and Martyr, and "fairest saint in all the calendar." It was her day, December 4. (On May 9, while the Twentieth Century Saint Barbara was in the midst of her martyrdom, Pope Paul VI declared Barbara a "dubious saint" and struck her from the calendar. The Santa Barbara Board of Supervisors noted wryly, in a resolution, that he had done this without so much as a public hearing; and that they, the supervisors, still considered Barbara saintly and would be honored to continue using her name in their city, county, channel—wherever it applied.)

In 1776, Father Pedro Font, a Franciscan monk, observed that ". . . much tar which the sea throws up is found on the shores. Little balls of fresh tar are also found. Perhaps there are springs of it which flow out of the sea."[2] This is one of the first recorded references to a natural oil seep.

By the time of Father Pedro Font's observations of the oil, the interest of the Spanish in California included not only the expansion of their empire but a genuine conviction that the heathen Indians needed to be "saved." Most famous of the early padre-explorers in the west was Father Junipero Serra, whose dream it was to spin a web of missions across California. In pursuit of his dream, on April 21, 1782, he founded the Mission Presidio of Santa Barbara, to serve the Canaliño Indians who by now numbered nearly 15,000. The mission was built and officially dedicated in 1786, becoming the tenth mission to be established in California.

Despite Spanish territorial claims, British exploration continued. References to the omnipresent oil continued, too. In 1792, Vancouver, famous navigator for the equally famous Captain Cook, recorded that:

> The surface of the sea, which was perfectly smooth and tranquil, was covered with a thick, slimy substance, which when separated or disturbed by a little agitation, became very luminous, whilst the light breeze, which came principally from the shore, brought with it a strong smell of tar, or some such resinous substance. The next morning the sea had the appearance of dissolved tar floating on its surface, which covered the sea in all directions within the limits of our view.[3]

When the revolt of Mexico against Spanish domination began in 1810, Santa Barbarans merely hauled down their Spanish flag and hoisted that of Mexico. But soon the Spanish supply ships stopped coming, California became isolated, and hard times fell upon the people. Then Yankee trading ships began to arrive in increasing numbers. At first they merely put in at the California ports to exchange cloth and implements for hides and tallow. But more and more Yankees came to stay and to intermarry with the local women. Besides the sailors, a second type of immigrant, the fur trapper, crossed the Rocky Mountains in pursuit of the sea otter. In 1840, Santa Barbara was a tiny cosmopolitan settlement of about 900, composed of a few French, a few Irish, a sprinkling of Old World Spanish, and the blended descendants of earlier arrivers—the beginnings of the American melting pot.

With the Mexican war of 1846 came the conquest and annexation of California to fulfill the "Manifest Destiny" of the United States. Santa Barbara began her legal existence as an American city four years later on April 9, 1850, when she was incorporated by an act of the first California Legislature. Slowly at first, then more rapidly, Santa Barbarans banded together to build and preserve the beauty of their city. Their tradition of civic pride and unity extends to today, while their ambivalence toward an influx of newcomers who bring both dollars and unwanted change was eloquently expressed by an editorial in the *Morning Press* of January 3, 1874:

> The old landmarks and most charming characteristics of Santa Barbara are disappearing before the march of "improvement," and though our practical people cannot move the mountains, nor change the scenes, nor spoil the climate, they are doing all they can to despoil the quaint beauty of the place and make it just a commonplace American town. . . .

How timely this sounds! The lament of Santa Barbara becomes even more poignant now, a century later, with our diabolical skill at doing those very things: moving mountains, changing scenes, and spoiling climates.

Nature and history have relentlessly bound oil with the destiny of Santa Barbara. Natural seeps are frequent along the margins of the channel and offshore in the tidal zone, and even on the deeper sea floor. Trained observers for the U.S. Coast and Geodetic Survey and the Commission of Fish and Fisheries have reported oil slicks in the channel since 1850. One investigator, in 1889, ". . . sailed through a most extraordinary region of the channel in which there is a submarine petroleum well. The surface for a considerable distance is covered with oil, which oozes up from sources below the water, and its odor is very marked."[4] It is easy to guess how Coal Oil Point got its name. Today, residents of the area will admit that throughout their memories, globs of oil have been frequent, if unwelcome, visitors to their beaches and water recreation areas.

Although many of the onshore asphalt deposits have been obliterated, archeologic evidence of them is indisputable and literary references abound. One of the largest asphalt deposits was at Carpinteria, about half a mile southeast of town near the sea cliff. By 1857, this deposit was supplying material from which illuminating oil was distilled, and quarry pits covering several acres had been dug as deep as 25 feet. The shales, according to State Geologist Whitney in 1865, "are black and highly bituminous where the outcrop strikes the sea . . . and large quantities of tarry asphaltum flow from them. For a mile or more along the shore the banks abound in it, and it saturates the beach sand and flows down into the sea."[5] Another large asphalt deposit was located near More's Landing. Whitney reported that this deposit was mined as a source of roofing and paving material. A similar deposit, near the coast at Elwood, consisted of veins of asphalt so thick and pure that they were mined to depths of 100 feet. A point on the coast half a mile east of the mine was heavily coated by petroleum washed ashore from offshore seeps. Old San Francisco waterproofed its roofs and paved its streets with tar from Santa Barbara.

A case might even be made for the theory that the very success of the old mission town of Santa Barbara as a tourist attraction was founded on the occurrence of natural oil seeps. By the 1880's, Santa Barbara was a small city of about 500. Her pleasant climate, coupled with her reputation as a health resort, had begun to attract visitors from near and far. Not only did they come from other parts of California on the new branch of the Southern Pacific Railroad, but they came from as far away as Boston. Attracted by grandiose advertisements published in eastern newspapers, they wanted to partake of the wondrous health cures promised to all. How old is the game of salesmanship! It was claimed that the prevailing southwest wind, blowing over a large and persistent oil slick on the ocean about six miles west of town, purified the air over Santa Barbara and not only made infectious diseases a rarity but also had a salutary effect on almost all chronic illnesses.

Oil Exploitation in Santa Barbara

With the discovery of oil in the Summerland district in 1895, systematic exploitation of oil in the area began. The discovery was made when, drilling for water some 500 feet back from the beach, the drillers struck first a vein of natural gas and then, deeper, a heavy grade of petroleum. The search for black gold was on. It led back under the ocean bed, as hungry oil hunters traced the extent of the producing sands. The Summerland Oil Company constructed long wharves into the sea, sank a conductor pipe to solid foundation, and bored the first oil wells ever to pierce the ocean floor. Quick to recognize a potential dollar, promoters for the Southern Pacific Railroad advertised their Los Angeles-Santa Barbara run with the slogan "See the Oil Wells in the Sea." The successsful drilling of the first well precipitated a scramble for water frontage, with the subsequent installation of more than 300 derricks by the end of the century. This was the third producing oil field in California. It proved to be shallow, however, and production dwindled rapidly. By now, heavy gales have swept away most of the old piers which once supported the oil rigs, but even today tourists flock to view the amputated stumps of piers still visible offshore about Summerland.

At the turn of the century, oil companies were buying up vast tracts of land in California. Activities of the Union Oil Company in Santa Barbara County go back to these early days of exploration, speculation, and development. In 1903, it bought the property on which stood the abandoned La Purisima Mission, and then offered the mission back to the county or state on the condition that it be restored as a historic shrine. Was this a conciliatory gesture in public relations? Perhaps in reciprocation for other considerations? If so, it is rather strange, because it was some years before public interest in the restoration project could be stimulated. Furthermore, most Californians were more than happy to welcome oil in those days, when the lives of entire towns depended on it. For whatever reasons, the land bearing the Purisima ruins was eventually deeded to Santa Barbara County, which passed it on to the state, in 1935, when actual restoration was begun.

The town of Orcutt, lying in the heart of the Santa Maria oil district in Santa Barbara County, owes its name and its existence to the Union Oil Company. It was a Union Oil geologist, W. W. Orcutt, who in 1902 first mapped and appraised the surface geology of the entire district, whereupon the company began buying land and leases. By 1904, Union Oil had brought its holdings to 70,000 acres, thereby controlling most of the oil pool. The Pacific Coast Railway built a spur track to the new oil field, the oil company itself put down 32 miles of 6-inch oil pipe along the railroad right-of-way, and in just one year the town of Orcutt became a thriving oil center.

The Union Oil Company also figured in one of California's earliest and most remarkable gushers. It was in 1904, in the slap-happy days of prospecting, when a gusher heralded the discovery of a rich, easily accessible oil pool and was cause for rejoicing rather than humiliation. Tremendous pressure forced oil up the 8-inch casing and 150 feet into the air from Union Oil's Hartnell No. 1, named after the Hartnell property on which it stood. For three spectacular months, Hartnell No. 1 spouted half a million gallons of oil a day, and it continued to flow vigorously for two more years before it could be put on the pump. It was one of the wonders of California, until it was surpassed by the Midway Field gushers of Kern County some six years later.

Production of oil on a large scale in Santa Barbara County began in the 1920's, at about the same time that the state of California introduced regulations governing offshore development and began to grant exploration permits and leases. The offshore parts of the Rincon field were discovered in 1927, the same year that saw Santa Barbara's boom ushered in by the discovery of the Elwood field. At its peak production of more than half a billion gallons per year, Elwood was the fifth highest producing field in the state. Subocean drilling, introduced at Summerland in the 1890's, was better developed at the Elwood field, where a dozen derrick piers were erected.

In 1929, in its repeal of the State Mineral Leasing Act of 1921, the California legislature prohibited new drilling in tideland areas. Although the new law did not apply to the existing Elwood wells, curtailment of activity by the drilling companies temporarily rendered further exploitation of the tideland areas unprofitable. The oil industry was largely confined to onshore activities until after World War II.

Shortly after World War II, new breakthroughs in petroleum technology encouraged several of the major oil companies to begin extensive geological and geophysical exploration of the offshore areas. They now had at their disposal the techniques of piston and jet coring of surficial deposits, as well as shallow and deep test drilling programs and seismic surveying. The state of California again began to grant permits and leases for oil exploration and production, and offshore activity expanded rapidly under the vigilant but approving eye of the state. A peak in the exploration program occurred in 1956-1959, when new methods and improved equipment made it feasible to drill to depths of 7500 feet and to operate in water as much as 400 feet deep. In 1958, Rincon Oil Island was completed 2800 feet offshore. In the same year, five additional tideland parcels were awarded by the state between Goleta Point and Point Concepcion, leading to rich new oil discoveries. So far California and the oil industry were happy partners in their enormously profitable venture.

But the citizens of Santa Barbara were not completely happy. Since the turn of the century, there had been intermittent complaints and organized protest from Southern Californians whenever beaches and coastal water were fouled by oil. Until the 1950's, these protests were almost always directed at the U.S. Navy or the shipping industry, which were accused of pumping bilges in the nearshore waters or in port. Both the Navy and the shippers did this, and frequently. But it is curious that, even though the existence of offshore oil seeps had been recognized for hundreds of years, they were apparently never considered as possible sources of the oil and tars that irate bathers found on their beaches. Perhaps human beings must always identify a human villain as the cause of any discomfort. Perhaps it is too frightening to feel helpless before chance or the power of natural forces. At any rate, since 1950, the oil industry has borne the brunt of the suspicion and criticism from Californians who are constantly fearful of losing their cherished water playgrounds.

Following the discovery of the Coal Oil Point offshore field in 1961 and the Alegria, Caliente, and Molino offshore fields in 1962, no new discovery was made in the tidelands for four years. Then, in 1966, the Carpinteria offshore field was discovered, which led to the first sale of a federal lease beyond the three-mile limit in the Santa Barbara Channel. Now a storm of protest broke loose, culminating in the Santa Barbara County Report of 1967. Nevertheless, since 1966, offshore developments have been almost exclusively on federal leases; and local citizens' groups battling to keep oil out have won only small concessions from the powerful oil interests and from the government, which covets the revenue from offshore leases. The particular lease whose development led to the Santa Barbara blowout dates back to February 6, 1968. On that day, at the Biltmore Hotel in Los Angeles, the Bureau of Land Management presided over a major lease sale. The atmosphere was one of secrecy and anticipation. As officials opened and announced the bids, there were gasps of surprise and cries of delight. It turned out to be a record lease sale (subsequently to be surpassed by bidding on Alaska's north slope, $900 million in 1969, and on the Gulf of Mexico, $850 million in December 1970). The 5400 acres in tract 402, under Federal OCS P-0241, added a bonus of more than $61 million to the federal treasury, a new high of $11,373 per acre. Seven months later, Platform A was set. Drilling from the platform began in January 1969. And on January 28, 1969, the fifth well being drilled blew.

III Techniques and Hazards of Drilling for Oil

Sarcastic Science, she would like to know,
In her complacent ministry of fear,
How we propose to get away from here
When she has made things so we have to go
Or be wiped out. Will she be asked to show
Us how by rocket we may hope to steer
To some star off there, say, a half light-year
Through temperature of absolute zero?
Why wait for Science to supply the how
When any amateur can tell it now?
The way to go away should be the same
As fifty million years ago we came–
If anyone remembers how that was.
I have a theory, but it hardly does.

Robert Frost

When Colonel Edwin Drake drilled the first oil well back in 1859, his drilling methods were very crude. From a derrick about 40 feet high, he suspended a heavy steel drilling bit on a wire cable. As the bit was raised and lowered, it pounded and pulverized the rocks below. Periodically, he removed the cuttings from the hole with a long bucket called a "bailer." That is all there was to it: brute force pounding and scooping until the hole broke through into the oil-bearing formations.

Colonel Drake's well was drilled to a total depth of 69.5 feet, a mere pinprick in the earth. Since that time, the total depths drilled have increased steadily with the passing years so that today many wells can be measured in miles, rather than in feet.

How to Drill an Oil Well

The modern rotary drilling rig was introduced around 1900 in the Spindlemont Field near Beaumont, Texas. The most conspicuous part of this rig is its derrick, which usually is between 100 and 200 feet tall. Instead of the bit's being attached to a wire cable, it is screwed onto heavy pipe called drill pipe, to which additional sections (stands) are added as drilling progresses. At the base of the derrick are the rotary table and other drilling machinery. As the rotary table rotates the drill pipe, the bit at the bottom of the pipe bores through the rocks.

A special fluid called drilling mud is pumped down through the drill pipe and bit, up through the space between the drill pipe and the wall of the hole which has been cut, and back to the surface where it is screened and tested before it is once again pumped back into the well in continuous circulation. This drilling mud serves several functions. It cools and lubricates the bit; it blasts the cuttings from the bottom of the hole and floats them to the surface; it seals the wall of the hole to prevent caving; and it serves as a hydrostatic counterpressure to keep any high pressure gas, oil, or water from blowing out of the hole uncontrolled.

Monitoring and adjusting the weight of the drilling mud is a critical and difficult operation. If the mud is too heavy, its pressure is too great and it can fracture a weak rock formation. Drilling mud can then escape into the fractured rocks, resulting in a condition called lost circulation. If, on the other hand, the mud is too light, its pressure is not great enough and drilling through a high-pressure oil or water formation can cause the well to blow out of control. The problem of finding the correct weight for the drilling mud is therefore compounded by the fact that as drilling progresses, strata of greatly differing strengths and pressures are successively encountered. A delicate balance is required, involving adjustments in mud density of as little as one ounce per gallon (one ounce out of about twelve pounds). This is a difficult feat for most people to accomplish on a small scale in a laboratory. A drill boss makes these fine adjustments on thousands of gallons of mud at a time, under often less-than-desirable field conditions. He is a very important and highly skilled man.

From time to time as drilling progresses, lengths of steel pipe called casing are set into the well to prevent cave-ins or blowouts and to prevent the entrance of gas, oil, or water into the well before it is completed. In order to hold the casing in place and to seal off a water or gas formation, it is customary to set the casing through and below the porous formation and well into solid rock. In offshore operations, the first length or string of casing is called conductor casing. The conductor casing extends from the platform at the surface of the sea to the sea floor and on down to solid rock, where it can be firmly anchored. It is usually inserted to a minimum depth of 300 feet, and the annulus between the casing and the wall of the hole is cemented solid for its entire length below the sea floor. In the case of Santa Barbara, Regional Oil and Gas Supervisor Don Solanas had told the Union Oil Company to install conductor casing in well A-21 to a depth of only 240 feet, because an oil sand was located at the customary 300 foot depth and drilling into that for the conductor casing would have opened the oil sand to the sea.

The next string of casing is called surface casing. Its length is determined by the ultimate depth of the well, and it varies from one quarter of the proposed total depth of a shallow well up to 2500 feet for a well two or more miles deep. Following the surface casing, strings of intermediate casing are set and cemented as required. The total number of intermediate casing strings is constrained by the requirement that each string be of smaller diameter than the one before, as each must be lowered through the part of the well that has already been cased. The final string, called the production casing, penetrates the lowest producing formation which that well will reach.

A completed well is put into production by perforating the walls of the casing opposite the desired oil or gas zone. Oil then flows from the pores in the formation into the well, driven by the energy of dissolved gases escaping from solution. If this natural reservoir energy is depleted while large amounts of oil still remain in the formation, various artificial means of driving the oil to the surface may be employed.

Unitization

Competition in the oil industry is intense. In the early days, many wells were drilled into the same formation by different operators, each eager to withdraw as much oil as he could at the expense of his competitors and, often, at the expense of the total productivity of the oil field. To minimize this destructive competition, holders of federal leases are encouraged to join together in "unitizing" lands on a prospective geologic structure for operation by a unit operator. Unitization affords such benefits as minimizing dissipation of reservoir energy and thus maximizing oil recovery from the formation, eliminating expensive competitive drilling, and allowing the most efficient exploitation of the physical characteristics peculiar to the reservoir in question. Unitization is the most practical, economic, and conservative method of developing an oil field. It is the basis on which the Union Oil Company and its partners were operating on tract P-0241 in the Santa Barbara Channel.

Sometimes There Are Accidents

In the early days of the oil industry, it was expected that any rich oil field would produce its share of "gushers"—oil wells gone wild, spewing thousands of barrels of oil high into the air for months and sometimes years before they could be tamed. The public attitude toward gushers, nurtured in the past by Hollywood, has varied from amused tolerance to enthusiastic acclamation. Two generations of movie-goers have thrilled at the sight of Clark Gable and Spencer Tracy hugging each other ecstatically, while uncounted gallons of wasted oil rained down on their heads. Often these spectacular black geysers were accompanied by equally spectacular fires that were almost as hard to control as the oil itself. In the movies, John Wayne, peerless fire fighter, always saved the day. In real life, however, the fires sometimes raged for months. Prodigious amounts of oil were wasted, property was destroyed, and, often, lives were lost in these (what were thought to be) inevitable accidents.

But creation of a gusher is not an act of God; it is an act of Man. Science and technology, advancing hand in hand, have rendered this sort of accident not only *not* inevitable but, in most cases, inexcusable. The public attitude, too, has changed. These accidents, these results of human error, are no longer admiringly called gushers, but are called–in horror and dismay–blowouts. A blowout is an uncontrolled flow of well fluids and (or) formation fluids–gas, oil, or water. The glory and romance of a gusher is gone; the devastation of a blowout remains. Speak softly, Hollywood, lest Santa Barbara hear. Your gusher is a blowout.

There is almost no reason for a blowout ever to occur. The oil industry weeps about the enormous economic loss it suffers each year from blowouts. But the fact is that the industry has carefully weighed the potential losses from chances and short cuts that it takes against the increased production costs it would incur if it made maximum provision to prevent blowouts. Apparently, some operators thought they had found a profitable compromise between safety and expense.

It is possible, through the use of high density drilling mud, to counteract almost any formation pressure. The heavier the mud is, however, the more expensive it is and the slower and more difficult drilling becomes. It is also possible to protect weak formations against fracturing by the installation of numerous casing strings and cemented liners. This, too, becomes expensive and is limited in practice by the number of successively smaller strings that it is feasible to use.

An alternative to the costly protection obtained from heavy drilling fluid and numerous casing strings is a drilling strategy which assumes some acceptable level of risk and makes provisions for handling trouble if it occurs. Each well is equipped with several blowout preventers and other safety valves and devices for closing off the well until the trouble can be corrected. Again, each emergency device adds to the cost of the well both because of the initial expense and because of the necessity for periodic maintenance and testing of the extra equipment.

A threatened blowout is called a kick. It is caused by an imbalance of pressures between the formation and the well, which permits entrance into the well bore of formation fluids. This imbalance of pressure may result from one of several situations. The density of the drilling fluid may be too low. This is usually a calculated risk allowed by the drill boss for the sake of economy. In haste or in carelessness, the crew may fail to keep the hole adequately filled with drilling mud. This has the same effect as using mud of low density, because the pressure in the well depends on the hydrostatic head, and thus on the depth of the mud column. The hydrostatic pressure may also be reduced by too-rapid withdrawal of the drill pipe, a phenomenon known as swabbing, which creates an effect something like suction. All of these situations are preventable.

Sometimes drilling mud is lost to a surrounding porous formation. This is especially likely to occur during drilling into an abnormally pressured permeable zone. When loss of drilling fluid or lost circulation occurs, there is no standard method of emergency procedure. The choice of procedure depends on the circumstances of the particular well and on the wisdom and experience of the drilling crew.

Kicks develop into blowouts for one or more of several reasons. The first is failure of early detection. Technology has solved this problem in some cases by providing automatic monitoring and warning devices. One of these indicates whether the mud is being added to the hole fast enough to replace the volume of pipe being withdrawn. Another, when it is installed on the mud outflow line, indicates whether gas is entering the well bore. At the time of the Santa Barbara blowout, however, not all companies were using these devices. It is not known whether the Union Oil Company was using them on well A-21 or not. The second cause for a kick to develop into a blowout is that the crew takes the wrong initial action. This obviously is entirely a matter of judgment and skill, and no third party can legislate regulations that prevent people from making stupid mistakes. The other reasons for a blowout concern the control equipment itself: it may be inadequate, or it may not function correctly when it is needed.

About half of all blowouts are caused by failure to keep the hole full of mud, and most of these occur as the drill pipe is being withdrawn. It is not enough to add mud to the hole at regular intervals and hope things will come out even. It must be known at all times whether formation fluids are entering the well bore and whether the volume of mud added equals the volume of pipe removed.

Just how does a blowout develop? Formation fluid (kick fluid), oil or water mixed with gas under pressure, begins to enter the well bore. It blends with the drilling mud circulating past the point of entry. Because the kick fluid is of lower density than the drilling mud, this blending increases the pressure differential between the well and the high pressure formation, and allows kick fluid to enter the well at an ever-increasing rate. Meanwhile, degassing of the mixture occurs and the gas bubbles expand as they move upward to levels of lower pressure. Because of the enormous volume expansion of the gases as they come out of solution and the accelerating inflow of kick fluid, explosive forces develop and the kick rapidly becomes a blowout. This is what happened on Platform A, when at first the drilling mud shot high into the air, to be followed within seconds by a jet of oil and gaseous hydrocarbons.

Is Santa Barbara a Risky Place to Drill?

Statements made to and by the press, and elsewhere, have alleged that the Santa Barbara Channel presents unique hazards to oil drilling. Two features of oil fields may increase the risk of accident by reducing the margin of error in drilling and producing the oil. The first of these is the solidity and thickness of the caprock, which determine the ease of cementing the conductor casing as well as the margin of safety for containment of shallow oil. The second is the existence of high pressure in oil-bearing formations, which requires special precautions. The caprock in the Dos Cuadras field is thin and weak, but not uniquely so. We would conclude (especially with the advantage of hindsight) that this feature might have been sufficient cause not to exploit the Dos Cuadras field. However, even if this conclusion had been reached it would hardly have satisfied the people of Santa Barbara, because this situation prevails only on the two leases which include the Dos Cuadras field. Elsewhere in the channel, caprock is adequate and no such objection exists (see Chapter II).

The question of pressure in the field is both more complicated and more widely misunderstood. There are physical limits to the reservoir oil pressures that can occur. Pressure cannot be less than hydrostatic pressure (the pressure exerted by a column of water of equal depth). If it should somehow become less, the entire column of rock and water would subside until the pressure was once again at least equal to hydrostatic. The maximum oil pressure that is physically possible is determined by the additional weight of the rocks, and the total pressure in an oil pool cannot exceed that at which the oil would break out of its confinement directly through the rocks. This maximum possible oil pressure is called the lithostatic pressure.

To answer the question of risk in drilling, one must ask, risky in comparison to what? As we have seen, on the large scales of distance involved, rocks have a very low rupture strength. Yet at the greatest depths to which we have drilled in the Dos Cuadras field, the pressures are only slightly above hydrostatic. The margin of safety appeared to be adequate. By comparison, some fields in the Gulf of Mexico and elsewhere hover dangerously near lithostatic pressure for much of their depth. Such fields can be and are exploited without accident, although great care is required. Small errors in the weight of drilling mud or other slips can be disastrous. Blowouts through the overlying rocks are extremely difficult to control. In terms of ambient pressures in the field, then, we must conclude that the Dos Cuadras field is not an especially risky place to drill. Indeed, being near hydrostatic pressure, it is toward the safe end of the risk scale.

But What About Earthquakes?

Of all the natural events of the physical world, none is more terrifying than a great earthquake. Abruptly, without warning, the stable reference of the solid earth vanishes. The actual damage done by earthquakes rarely matches that due to seasonal hurricanes or floods, yet in the minds of people who have experienced a violent earthquake there lurks the haunting memory of a world gone mad.

From the time the earthquake history of the Santa Barbara Channel was made known after the blowout at Platform A, an aura of incipient disaster cloaked every discussion of the situation. There was, at times, the clear implication that the earth could crack open along new or existing fault breaks and oil would flow out uncontrolled and uncontrollable. No one can guarantee that this will not happen, any more than he can guarantee that the sun will rise each morning. Yet in both cases we can look at the past for clues to the future. Throughout the millennia of recorded history, oil has lain in the pools but so recently tapped by man. In Alaska, the Middle East, Indonesia–yes, and California–these oil pools are in regions of intense earthquake activity. In all of this history, there is not one single case of an earthquake followed by massive oil release. Despite earthquakes of all sizes in and near oil reservoirs, the earth has never been induced to give up her oil by this mechanism. It seems unlikely that some future event will so rudely contradict the past.

But man has intervened with his insatiable thirst for energy. Does the addition of oil rigs and oil operations introduce a new risk of oil outbreaks from earthquake damage? Several sources of risk can be imagined: the platforms, and the pipelines to shore, the shore storage facilities, and the producing wells themselves. It was noted in the second report of the President's Panel on Oil Spills that the standards for earthquake resistance for offshore platforms were out of date and should be reviewed in the light of more recent knowledge. What the report did not say was that the required design standards for *storm resistance* (100-mile-per-hour winds, 40-foot waves, and so on) define a structure that exceeds in strength the requirements–new or old–of earthquake resistant design. Standards should be reviewed, for we cannot count on such luck the next time; but existing platforms *are* adequate so far as both storms and earthquakes are concerned. No platform has ever been destroyed by an earthquake, not in California, not anywhere else.

Underwater pipelines are less vulnerable than would at first appear. Steel is elastic. (If you doubt this, drop a ball bearing on a cement surface and watch it bounce.) On the scale of an 18-inch pipeline several miles long, this elasticity can accommodate considerable fault displacement provided the pipeline is not rigidly attached to the fault itself. Hundreds of pipelines, not all of them for gas and oil, crisscross fault zones in California and withstand earthquakes quite well. Should an accident occur, there is a ready technical safeguard. Automatic shutoff valves, placed strategically along the pipeline, would allow only a small amount of material to escape in the event of a break from whatever cause. Such devices add expense, and oil companies prefer to use only a few. A militant public and its government could insist on a closer spacing of automatic shutoff valves than present practice provides.

Onshore oil storage facilities have experienced severe damage during large earthquakes. In the 1964 Alaska earthquake, to cite one example, damaged storage tanks led to raging fires at Seward and at several other locations. But these tanks were not related to offshore production. We have oil storage facilities wherever we have cities, because we demand oil. So long as we demand large and ready reserves of oil and gas, we need better designs and safeguards for our storage facilities, for the resistance of the present structures to storm and earthquake damage is low. But let us also look in *general* at our standards for earthquake-resistant design. The collapse of hospitals, schools, stores, and homes during recent earthquakes suggests that only luck has prevented some gruesome disasters. We persistently ignore this evidence at the peril of our lives.

The producing wells—what of them? Omnipresent in the concern following the Santa Barbara spill was the idea that the producing wells would open up and spew the contents of the pools they tapped into the sea. We have less history to go on for this case, but it all points in the same direction. Producing oil wells both onshore and offshore have experienced earthquakes without any resultant spillage or leakage. The Long Beach offshore field is one whose wells have sustained earthquake damage. The results there and elsewhere, as might be expected, were that the wells were sealed off as faults shifted. To scale, wells are tiny pinholes in the earth. As a fault moves or the earth slides and slumps, the problem is to keep these pinholes open. Earth tends to heal, not exacerbate, these punctures.

But to visualize the overall risk from offshore drilling, let us look at history in a different way. Let us project our past experiences into the future. We all use oil in one way or another, and if our growing population persists in demanding oil in increasing amounts we (and our environment) must pay a price. In the second report of the President's Panel on Oil Spills, we find a statement of what we may expect, based on a projection of past experience into the future:[1]

> In the case of offshore oil and gas we now have sufficient experience to make some comment. Since 1954, approximately 8000 offshore wells have been drilled. 25 blowouts have occurred of which 17 leaked gas only. Two resulted in serious oil pollution incidents and 9 constituted serious blowouts that persisted for several days with fires (9 cases) or fire hazards and hazards to personnel (29 deaths). If offshore development continues to expand at the present rate and the frequency of accidents remains the same, 3000 to 5000 wells will be drilled annually by 1980 and we can expect to have a major pollution incident somewhere every year.

More recent events like the two major oil spills on the Gulf Coast suggest that this estimate was, if anything, too conservative. We have had more than one incident per year already.

Technology can help some in producing safer operations, but there is no safeguard against human stupidity or carelessness. If production offshore continues to increase, accidents will also continue. Who shall pay the price?

IV Management and Mismanagement of Oil in California

There is something I don't know
that I am supposed to know.
I don't know what *it is I don't know,*
and yet am supposed to know,
and I feel I look stupid
if I seem both not to know it
and not know what *it is I don't know.*
R. D. Laing

Much of our national and individual behavior seems predicated on the belief that our founding fathers thought of and provided for every future situation that might arise. Sometimes our faith is even more sublime. We trust that, before America was born, the British thought of everything; or the Romans, before the British. When stated in this way, of course, the idea is nonsense, for there *are* new things under the sun. Events have led us many times into situations of which the wisest seers could not possibly have dreamed, and hence they could not have laid foundations for sound decisions at a later day.

One of the events which no one in the 1780's foresaw was the explosive development of the oil industry, with its concomitant jurisdictional and regulatory problems. Oil had been known since the earliest days of the colonies. Sometimes it was sought out in its natural seeps for its medicinal or lubricating properties. At other times, when it fouled beaches or fresh water supplies, it was a nuisance. There were spasmodic attempts at mass production and refining of oil, but for the most part the demand was low and easily met, and no one dreamed of the vast pools of black gold that lay under both the land and the sea, waiting to be exploited by future energy-hungry generations. When lawmakers of the early days thought of mineral wealth, they thought of gold and coal, not of oil.

By the 1830's, the demand for oil had risen sharply, largely because of its increasing use as an illuminant. In the 1850's, companies began to organize for the production of petroleum products. In 1859 at Titusville, Pennsylvania, Colonel E. L. Drake supervised the historic drilling of the first oil well. The year 1865 saw California's first oil boom—a short-lived affair, which ended in disaster. But the oil mania had been born, and neither the industry nor the government was prepared to deal with it.

Who Owns Oil?

Who had title to resources lying under the land? Historically and legally, it was not clear. The original thirteen colonies appeared to have received rights to their own mineral wealth directly from the British Crown in the Treaty of 1783. The situation in California, however, was different. Under first Spanish and then Mexican rule, mineral resources remained the property of the government, with title only to the surface of the land passing to individual holders. Ostensibly, then, in 1848 when the United States as a nation signed the treaty of Guadalupe Hidalgo, mineral rights passed to it rather than to the new state of California. Yet there was no explicit agreement on this matter, and California's interests were protected by a law insuring that no state after the original thirteen would be admitted to the Union under any agreement that put it at a disadvantage relative to the original states.

Without any decisive tests or definitive agreements, mining interests had essentially free rein for more than a century. Federal authority was weak in far-off California, and federal mining policy uncertain. It was an accident of history that the gold rush and the first oil boom came at a crucial stage in the development of a land policy, after three quarters of a century of effort following the Revolution had failed to produce a workable plan for disposal of United States mineral properties. When Congress finally got around to passing its Act of July 26, 1886, it did little more than legalize local mining customs already established by miners in the interests of miners. The issue did not go undebated, however. Many Congressmen felt that miners should buy or lease their mineral lands to help pay for Civil War debts.

California's Black Gold

In 1860, the state of California, eager to exploit its mineral wealth, authorized a state geological survey and appointed Josiah Whitney as State Geologist. Whitney envisioned his job as the scrupulous supervision of a thorough, scientifically respectable survey of California, including botanical and zoological observations. The legislature, however, had different motives. It appropriated money, provided the necessary staff, and then sat back to await discoveries which would bring riches to the state.

The legislature was already becoming disenchanted with Whitney and his project (a botanical survey? a *botanical* survey!) when Benjamin Silliman, Jr., arrived on the scene. Silliman was professor of chemistry at Yale and consultant to oil men in Pennsylvania and California. As President of the Pennsylvania Rock Oil Company, he had supported and encouraged Pennsylvania's neonatal oil industry until Drake's well vindicated his faith. Oil speculators in California were naturally eager to receive Silliman's blessing. He arrived in Santa Barbara on June 25, 1864 and, impressed by the natural seeps in the area, declared that as much oil ran into the sea and was wasted every day in Santa Barbara as was produced in all the rest of the country. In his typical hyperbole he described a natural seep off Goleta:

> Often for hundreds of acres square at one view, there is no part of the sea but was thus covered. The sea boils like effervescing soda water, with the escaping gas which accompanies the oil, and great globules of pure oil rising with the gas flash out on the surface of the water, tossing it up in jets, and then breaking into films of rainbow hues like the tints of a dying dolphin. The effect is wonderfully beautiful and exciting, every fantastic form of animal, plant, and fish is reproduced on this marbled surface in thin films of fine oil.[1]

On the basis of hasty exploration and analysis of a few samples provided by the oil companies, he made extravagant claims about the oil reserves of the state. Of some property at Ojai, he wrote that it ". . . covers an area of 18,000 acres in one body, on which there are at present at least twenty natural oil wells, some of them of the largest size. The oil is struggling to the surface at every available point, and is running away down the rivers for miles and miles."[2]

Silliman summarized his attitude toward oil in the report of his observations at Ojai Ranch:

> Again, every mine of metals is a magazine of limited supply—you own so many feet, and this area contains so much and no more—every dollar's worth of ore mined and sent away leaves one dollar less in store. Not so with petroleum. It flows on year after year, and still the source of supply seems unimpaired; no conflicting claimants can arrest its flow, and like the widow's cruise, the daily modicum is never wanting.[3]

Silliman's enthusiastic promises were exactly what the avaricious oil-seekers longed to hear. Thus, almost singlehandedly, one man precipitated the California oil boom of 1865. It ended, a few years later, when those first-discovered shallow oil fields were pumped dry, a disaster brought on by the state's obsession with economic gain and by the personal greed of the flamboyant Silliman and ruthless oil operators. Whitney had warned of the risks of inflated drilling ventures. The events of the 1860's brought disrepute to Silliman and temporary vindication to Whitney and his Survey.

The turn of the century witnessed a new oil boom in California. It was a malignant cycle of frenzied drilling, overproduction, criminal waste, falling prices, and extensive campaigns to increase consumption. The price of oil plummeted from a dollar per barrel to ten cents, forcing many small companies out of business. The insane response of the large companies was to drill even more wells and to force more and more of their product on an already glutted market. Like lemmings dashing heedlessly to the sea, they seemed intent on a suicidal race to drain every last drop of oil from the earth. Wastage was profligate—why conserve a product that could not be given away?

California was quick to recognize the problem of oil conservation, for at the same time that Silliman's predictions were apparently being fulfilled, other geologists were predicting the exhaustion of this irreplaceable resource. As early as 1903, California adopted a law to regulate the casing and plugging of wells. Many other laws followed, but it was not until 1915 that the state established a special division with a State Oil and Gas Supervisor to administer state regulations.

There was an oil shortage during World War I, as, again, fields were pumped dry and the demand for oil increased. Then, with developing technology and in the wake of the federal lease law of 1920 and the California lease law of 1921, still another California boom was ushered in. It, too, was characterized by overproduction, waste, overflowing reservoirs, declining prices, and frantic efforts to stimulate unimportant uses of oil, amidst predictions that supplies would run out. Still, both state and federal governments seemed reluctant to take the drastic steps needed to bring the situation under control.

What About Offshore Oil? The First Lease Laws[4]

About this time, another problem appeared. If it was not clear who controlled oil reserves under the land, the status of oil under the sea was even more confusing. It was generally assumed that the land between high and low water (the technical interpretation of "tidelands") belonged to the adjacent states. In addition, most states believed that they also controlled the sea from low tidewater to the three-mile limit recognized by international law. The original coastal states assumed that they had received these rights directly from the British Crown as provided in the Treaty of 1783. Later states claimed them under the "equal footing" law. In practice, however, the question of ownership of coastal water and the underlying lands and resources had never been answered because it had never been asked. Until it seemed both feasible and advantageous to drill offshore for oil, there had been no problems either actual or foreseen to force a decision. In fact, so academic did the problem seem that the coastline itself had been left undefined.

Despite the uncertain legal situation, both the federal government and the states had tacitly behaved as if the coastal states controlled their tideland areas and submerged lands out to the three-mile limit. California's constitution of 1849 described the area of the state as including islands, harbors, and bays, and having a western boundary line that was offshore "three English miles." Since 1872, California's civil code has declared the state to be owner of all land below low tidewater and all land below water of navigable lakes or streams. On the basis of these documents, California has regulated fishing and leased areas of kelp beds in its coastal waters, and made grants to individuals and local governments for various tidelands projects.

In 1920, Congress passed the Federal Mineral Leasing Act, which permits any person to file an application for an oil lease on public lands. California followed with its own lease law in 1921. Almost immediately, the state began to grant permits to private individuals or corporations to prospect for oil and gas in the tide and submerged lands. It also entered into leases of submerged oil areas in the Summerland and Rincon fields of Ventura County and the Elwood, El Capitan, Carpinteria, and Goleta fields of Santa Barbara County. Dollars from royalties began to roll into the state treasury. Administration of the law, however, developed many problems. For one thing, there was no provision for supervision of offshore drilling. For another, the Act left unclear the extent of the Surveyor General's powers to grant or withhold leases. When the development of the El Capitan area in 1926 brought an epidemic of several hundred applications, the Surveyor General decided to reject most of them. Angry oil men brought their grievances to court with the result that, in the 1928 case of *Boone* v. *Kingsbury,* the court ruled that the behavior of the Surveyor General was unconstitutional since the proposed leases were detrimental neither to fish nor to navigation. In effect, then, the decision threatened to undermine the entire state policy of leasing by allowing virtually unlimited drilling and requiring the Surveyor General to issue a permit for every application. It made a mockery of attempts to control the oil industry.

In response, in January 1929, the legislature called an emergency moratorium on further leases and a few months later it repealed the Lease Act of 1921. Some 350 wells had been drilled under the law.

In an attempt to bring together the scattered bits of its land policy, the state of California passed the State Lands Act in 1938. This was a comprehensive statute which placed all state lands under a State Lands Commission. Once more, the state began to issue leases or easements for oil exploration and production, but this time it was a conservation measure as well as a leasing act. It called for the granting of leases on the basis of competitive bidding, and allowed extraction of oil and gas from the tidelands only when state-owned oil was already being drained from existing onshore wells. For by 1938, technology had advanced to the point where slant drilling was profitable, and many such directed wells were pumping oil from under adjacent territory. Like the drillers, the oil pools did not necessarily respect political boundaries. They frequently underlay not only areas specified in a lease but also adjoining properties. As a result, strategic placing and slanting of wells could yield the clever oil man much more oil than he had officially bargained for. Sometimes too, because technology was not perfect, wells were accidentally slanted. For all of these reasons, oil was illegally being withdrawn from state lands, and the state acted to collect royalties on oil so removed.

The Tidelands Controversy

By the time of Franklin Roosevelt's second term, federal, state, and local governments were vying with one another for revenue from the leasing of their lands to the oil industry. The federal Mineral Leasing Act vested authority for granting leases in the Department of the Interior. For fourteen years, until 1934, the Department had routinely rejected all applications for leases in the California tidelands. Clearly, the federal government believed these areas to be under control of the state. Then, in 1934, Joseph Cunningham applied for a federal prospecting permit in offshore areas in California. Like others before and after, Cunningham's application was rejected. But Cunningham's group chose to press its demands before Congress. Thus the tidelands issue was born. Although most of the controversy was to revolve around the oil issue, the much larger question of federal versus states' rights was actually at stake.

The disposition of the offshore tidelands simmered as an issue in Congress for eleven years, accompanied by several preliminary skirmishes in the courts. Meanwhile, sentiment in Washington shifted in favor of federal control of offshore oil. The federal courts, which earlier had thrown submerged lands cases out of court for want of a federal issue, began to consider these suits. Conservation fervor gripped some members of Congress (notably those who did not come from oil-producing states), along with a desire to assert federal "rights" to the marginal sea. Several resolutions were introduced to establish this authority, but the majority of senators and representatives favored state control of the lands in question. It is noteworthy that, from outside the government, the only voices raised in support of federal control were from those who stood to reap financial gains from federal leases. Nevertheless, in the executive branch, Secretary of the Interior Harold Ickes shifted his early position favoring state control of submerged lands to one favoring federal control. He was supported by the Secretary of the Navy.

World War II relegated the tidelands issue to the background for a time, but it was not forgotten. If anything, concern for the national defense convinced Ickes more firmly than ever that the tidelands belonged in the federal domain. So with the death of Franklin Roosevelt and the end of the war, Ickes resurrected the problem and made a national issue of it. On September 28, 1945, President Truman issued a proclamation announcing that the United States government regarded all resources of the subsoil and sea bed of the continental shelf as federal property. Although this was designed mainly to clarify the U.S. position in international law, its implications for the states were clear.

On October 9, 1945, Attorney General Tom Clark filed suit against the state of California in the Supreme Court of the United States. This was the historic case of *United States* v. *California*. Its purpose was to settle, for all time, the issue of rights in the marginal sea. Two other suits were to be filed later, *United States* v. *Texas* and *United States* v. *Louisiana*, but these cases just replayed the arguments of the first. The claim of the United States was, in the words of Attorney General Clark, that it was the owner of or possessor of paramount rights in and powers over "the lands, minerals, and other things of value underlying the Pacific Ocean, lying seaward of the ordinary low water mark on the coast of California and outside the inland waters of the state, extending seaward three nautical miles and bounded on the north and south by the northern and southern boundaries of the state." The hope was for a declaration of the rights of the United States, and an injunction against the state of California and persons acting under it from trespassing on the federal territory. The main arguments advanced for the claim were that the United States needed jurisdiction over the marginal seas because it was in the best interest of the national defense, and because it was necessary in order for the United States to fulfill its role in international legal matters.

In June 1947 the Supreme Court rendered its precedent-shattering decision that the federal government indeed had paramount rights over land, minerals, and other resources in the disputed area. California immediately petitioned for rehearing and filed several suits to determine exact boundaries. Meanwhile, revenues derived from state leases offshore were held in escrow, pending settlement of the issue.

In 1952, a temporary compromise boundary was agreed upon. Then, in the Channel Islands decision of May 17, 1965, the coastline along the entire west coast was finally established. This decision cost California many of its claims, including those parts of the Santa Barbara Channel between the three-mile limit off the coast and the three-mile zone surrounding the four Channel Islands. It paved the way for the approaching controversy over federal leases in the Santa Barbara Channel.

Rights to the marginal sea became a national issue in the 1952 election. Those favoring federal control tried to limit the controversy to offshore oil, to gain support of inland and non-oil producing states. Eisenhower emphasized that what really was involved was states' rights, which he strongly supported. With the election of Dwight Eisenhower to the presidency, then, the tide was turned. Knowing that the new president favored formal surrender of federal claims to the continental shelf, Congress quickly passed two important measures. The first was the Submerged Lands Act of May 22, 1953, which deeded title and ownership of all submerged lands within the three-mile limit to the states. The second, which followed within months, was the Outer Continental Shelf Act, which designated the area outside state limits as the Outer Continental Shelf (OCS) and placed it under exclusive federal jurisdiction. This act did not extend the territorial boundaries of the nation, but limited its authority to the sea bed and subsoil. The seaward limit of the OCS remains undefined. In practice, our national claim may extend as far as exploitation of natural resources is possible—an ever-expanding boundary, which may invite conflict in the future. The Secretary of the Interior was authorized to lease areas on the OCS through competitive bidding. As a national issue, the question apparently was settled. California's agony was yet to come.

Oil and Santa Barbara

New methods and improved equipment led to a peak in offshore exploration between 1956 and 1959. Once more, the state began to issue leases for offshore drilling. Exploration, leasing, and development continued until by 1966 the state of California had leased all submerged land within the three-mile limit between the Ventura County line on the east and Point Concepcion on the west, except for a 16-mile sanctuary adjacent to Montecito, Santa Barbara, and the Goleta Valley. This was not accomplished, however, without anguished opposition from property owners, conservationists, and local public officials. In fact, it was the relentless protest of these groups that won for them the offshore state sanctuary where drilling was prohibited.

Since 1966, offshore developments in the Santa Barbara Channel have been on federal leases. The first federal lease sale on the Pacific coast took place in May 1963, when 58 tracts, 312,974 acres off the coast of northern California, were sold to four oil companies for $12,807,837. This sale went all but unnoticed by the southern California newspapers, which were preoccupied with the riots in Birmingham, Alabama, and the 22-orbit space flight of astronaut Gordon Cooper. Similarly, the 1964 lease sale off Washington and Oregon seemed of little concern to southern Californians. However, on November 3, 1965, Interior Secretary Stewart Udall authorized core drilling in the OCS lands between Point Concepcion and San Diego in three general areas described as the Santa Barbara Channel area, the San Pedro Bay area, and the Santa Monica Bay area. The U.S.G.S. would grant permits for core hole drilling in order that the oil companies might obtain adequate geologic data to promote maximum bidding competition at a future lease sale. Now southern Californians united in protest. Led by the Sierra Club, their outcry was loud and long.

In 1966, one federal lease was sold, a small tract of about 2000 acres near Carpinteria. This "drainage sale," which brought a bonus of $21 million and a first year's rental fee of $10 thousand, was necessary because the development of the newly opened Carpinteria field on state leases was subjecting the federal lands to drainage. The Standard Oil Company, which owned the adjacent state lease tract, had strategically located its drilling platform near the boundary of the federal land, in an excellent position from which to drill slanted wells and poach on the federal oil reserves. Concerned voices in Santa Barbara County warned that this would happen again and again–a domino effect, with each lease sale leading on to the next because of this kind of drainage problem, until the precariously protected buffer zone was eroded away.

After the flurry in 1966, the oil companies and the Interior Department could not forget the rich oil lands in the Santa Barbara Channel, and another lease sale was scheduled for late in 1967. Again, the announcement unleashed a storm of protest from the same groups that had contested the state leasing and had later opposed the federal sale of 1966. They won the first round of their battle on September 22, 1967, at a meeting in the Conference Room of the Santa Barbara County Board of Supervisors. At this meeting, Assistant Interior Department Secretary Harry R. Anderson agreed to reserve a two-mile "buffer zone" in the federal territory bordering on Santa Barbara's sanctuary, and granted a 60-day moratorium (they had asked for six months) for further study and evaluation of the effects of drilling in the channel.

This investigation culminated in a five-pound report prepared by the Oil Well Inspection Department of Santa Barbara County and submitted to the Board of Supervisors on November 15. It was a comprehensive statement of the geological setting of the channel, present and potential drilling platform design and technology, oil and gas economics, aesthetics, multiland use, regulations, pollution, and public opinion. In general, the report favored cautious development of the channel region, in a manner that would safeguard all of the area's resources, including oil itself. It rejected as "absurd" the Santa Barbara *News-Press* projections of 70 platforms in the eight-mile stretch outside the sanctuary, when 12 years of development along 50 miles of Santa Barbara County coastline had yielded only eight platforms to date. It was skeptical of the economic potential of the area—an attitude apparently not shared by Union Oil and its associates, and certainly not confirmed by subsequent events.

After receiving the report, the Interior Department announced ". . . we feel that maximum provision has been made for the local environment and that further delay in the lease sale would not be consistent with national interest or regional economic welfare." The bid call went out for November 27, 1967. Bids were opened on February 6, 1968. As we have seen, it was a record-breaking sale.

The government holds certain specific powers over offshore oil operations. Under an 1899 act of Congress, the War Department was charged with determining whether the construction of offshore structures in navigable waters interfered with navigation or commerce. Later this duty was vested in the Corps of Engineers in the Department of Defense. Not only every federal, but every state lessee since 1899 has advised the War (or Defense) Department of his lease and has received permission to erect each oil rig. In 1967, agitated Santa Barbarans pleaded with the Corps to step outside their narrow criteria for deciding for or against the oil rigs. They wanted the oil rigs to be disallowed because they, the people of Santa Barbara, did not *want* oil rigs in their channel. Like a telephone recording, the answer came back again and again: "The proposed structures pose no threat to navigation . . . the proposed structures pose no threat to navigation . . . the proposed structures pose no threat. . . ." The actual leasing of public lands is handled by the Bureau of Land Management through its State Field Offices, but the job of supervising and coordinating the development of federal oil and gas leases is delegated to the Branch of Oil and Gas Operations in the Conservation Division of the U.S.G.S., Department of the Interior. This organization, or its forerunner, was established a few years after the Federal Lease Act was passed, as the professional and scientific representative of the federal government in dealing with what is now more than 100,000 oil and gas leases on 68 million acres of public land. More than 1000 of these leases are on the Outer Continental Shelf.

There are seven regions in the United States under the Oil and Gas Branch. It was Don Solanas, Supervisor of the Pacific Region, who was summoned from his business in Wyoming to take charge when the trouble began on Platform A—Solanas, along with District Engineer Harry Cypher, who headed one of the three districts in the Pacific Region. The routine job of these people is to assure that all drilling and operating procedures are carried out in consonance with the leasing and operating regulations specified by the Department of the Interior. The regulations cover drilling and abandonment of wells, well-spacing and well-casing, necessary samples and tests, maintenance of well records, periodic reports, collection of rentals and royalties, and responsibility for accidents and fires. What disturbed Secretary Hickel as he viewed the oil spill was that the federal regulations had not been reviewed or updated in the 15 years since the passage of the OCS Act. Californians cried "It never would have happened if" federal regulations and their enforcement were as strict as those of the state. What they overlooked was the potential for human error and misjudgment.

Santa Barbarans have a love-hate relationship with the oil industry. In 1965, Santa Barbara County was the third ranking county in the state in production of minerals. The major sources of income in the county in 1965 break down like this:[5]

oil and gas	$98,000,000
agriculture	79,000,000
manufacturing	64,000,000
tourism	41,000,000
other minerals	28,000,000
commercial fishing	412,000

The tax levy for Santa Barbara County in 1967-68 was more than $52 million. On an industry basis, the oil industry was the largest taxpayer, paying almost 10 percent of this amount, more than $5 million. Two percent of the taxes came from offshore oil. Of the five largest taxpayers in the county, three were oil companies; and of the twenty largest, nine were oil companies. Clearly, Santa Barbara has much to gain economically from oil.

In the ten years following the first offshore drilling in 1957, federal, state, and county governments received more than $240 million from oil industry offshore operations as taxes, royalties, and bonuses. During the same time, state offshore leases paid $6.5 million in taxes to Santa Barbara County. This is a rate of 17¢ per barrel for the 37,569,000 barrels produced, or more than $1.6 million each year from offshore oil production.

In our usual American way of thinking, any enterprise which brings so much wealth cannot be totally despised. Santa Barbarans love their cars and boats and power lawn mowers as much as anyone else does. They can also afford more of them than most people can. But they love their beaches and wildlife as well. Get oil out, they say. Drill for it somewhere else, anywhere else, but not in the Santa Barbara Channel. Can they, or anyone, expect to have it both ways?

V Everyone Gets into the Act

The last word on how we may live or die
Rests today with such quiet
Men, working too hard in rooms that are too big,
Reducing to figures
What is the matter, what is to be done.
A neat little luncheon
Of sandwiches is brought to each on a tray,
Nourishment they are able
To take with one hand without looking up
From papers a couple
Of secretaries are needed to file,
From problems no smiling
Can dismiss; . . . far into the night
Their windows burn brightly
And, behind their backs bent over some report,
On every quarter,
For ever like a god or a disease
There on the earth the reason
In all its aspects why they are tired, the weak,
The inattentive, seeking
Someone to blame; if, to recuperate
They go a-playing, their greatness
Encounters the bow of the chef or the glance
Of the ballet-dancer
Who cannot be ruined by any master's fall. . . . But then
Perhaps one should mention
Also what must be a comfort as they guess
In times like the present

When guesses can prove so fatally wrong,
The fact of belonging
To the very select indeed, to those
For whom just supposing
They do, there will be places on the last
Plane out of disaster.
No; no one is really sorry for their
Heavy gait and careworn
Look, nor would they thank you if you said you were.
W. H. Auden

There was an anxious period of self-examination and re-examination at all governmental levels which began as soon as news of the oil leak was received. In Sacramento, Governor Ronald Reagan immediately ordered all state agencies involved with offshore leases to review their regulations, to assure that no similar incident could occur on state-leased lands. At the same time, Representative Charles Teague told the press that Secretary Hickel was already reviewing the federal regulations and that he suspected that they were not as strict as they should be. Senator Edmund Muskie's Air and Water Pollution Subcommittee hearings, held in the first days of February, attracted fifty newsmen and representatives of three major broadcasting networks. It was more coverage, boasted the Santa Barbara *News-Press*, than had been accorded to any other hearing except perhaps for Senator Fulbright's 1968 Vietnam hearings. Santa Barbarans were dissatisfied, however. They felt that their immediate plight was being overlooked in the generalizations and long-range planning.

The Santa Barbara County Board of Supervisors, led by George Clyde (perhaps the most vociferous opponent of oil on the Board), adopted a resolution urging cessation of all drilling for oil in the Santa Barbara Channel, both in state and federal waters. The Board was enthusiastically supported in this move by the Chamber of Commerce and other citizens' groups. Concurrently, the Carpinteria City Council was adopting a resolution requesting a ban on all new drilling and new platforms in the channel.

The House Public Works Committee, which had been holding flood control hearings in Los Angeles, came to Santa Barbara on Friday, February 14, to tour the disaster area and to hear testimony from invited speakers. The next week, a subcommittee of the state Senate met in Sacramento to consider the oil spill problem. And the week following that, on February 24 and 25, Senator Ed Muskie's Public Works Subcommittee on Air and Water Pollution met again in Santa Barbara.

Interior Does Its Homework

Meanwhile, the Department of the Interior was continuing its critical review of all regulations covering offshore drilling. Already, the study had revealed not only that there had been substantial disagreements within the Interior Department as to the adequacy of the regulations when they were being prepared but also that some people in the Department had questioned the wisdom of drilling in the Santa Barbara Channel in the first place. Udall himself, who ordinarily took top honors in the conservation business, had vacillated in his attitude toward the channel leases. He was later to refer to his decision here as a conservation Bay of Pigs. Regardless of what lay behind Udall's ultimate decision to go ahead with the channel lease sale, public announcement of the fact that his advisors had not been in agreement did little to comfort the Santa Barbarans.

Some of the language in the federal regulations that were in effect at the time of the blowout was rather vague. It admonished the oil operators to carry out their programs in such a way as to avoid accidents—like a mother advising a child to "be careful, don't take any chances." The new regulations left less to the imagination. As a first measure in revising the regulations, Hickel, on February 17, 1969, signed an amendment governing the liability of oil companies for pollution resulting from their activities. In effect, this amendment assumes negligence on the part of the operator, whereas formerly the burden was on the government to prove negligence. The next day, February 18, Hickel issued tentative guidelines for regulations "far tougher, far more modern, and far more in tune with the needs of all Americans."

Some of the more significant of the new regulations are as follows:

(1) Specifications for drive casing are given. The function of the drive casing is to provide stability for the initial drilling operations.

(2) The depth to which surface casing is required is increased.

(3) There is to be no more cutting corners on drilling mud; mud of the proper density must be used.

(4) A plan of action for emergencies must be submitted to and approved by the Geological Survey.

(5) Any major variance from the new casing requirements must be submitted to the U.S.G.S. headquarters for approval.

(6) Tests to ensure isolation of zones of oil, gas, and fresh water from each other must be witnessed by U.S.G.S. representatives.

(7) Blowout prevention requirements are more stringent.

(8) Warning devices are required to indicate whether the drilling mud in the hole is adequate.

(9) Safety and antipollution devices are required on platforms. These include gas detector and alarm systems, firefighting systems, shutdown devices for the wells and equipment on the platform, pipeline alarm, and auxiliary power equipment for safety. Pollution control equipment including booms, skimmers, and approved chemical dispersants, must be on hand during all mobile platform and drilling operations.

(10) Additional scheduled and unscheduled inspections of all operations will be conducted by the U.S.G.S.

What Is An Expert?

In the final revision of his department's regulations, Hickel and his colleagues relied heavily on suggestions made by oil company representatives. At first, this might seem to justify a cry of "Foul! The public is being screwed again!" But it would be wise to consider for a moment, who can best advise on technical matters? An expert? A second question, then, related to the first: what is an expert? These questions are of importance both for the revision of the drilling regulations and for the selection of members of the President's Advisory Panels on the oil spill.

If we assume that it is wise to seek advice from those who know most about the problem—from the "experts"—we immediately encounter the issue of "objectivity" and "conflict of interests." An expert in matters pertaining to oil is someone who is familiar with the scientific, technical, and perhaps economic and ecologic aspects of oil production. He knows the geology, physics, and engineering skills fundamental to the industry; and he is familiar with the hazards involved at each stage of production and with the current ability to cope with them. Where would you look for such a person? In one of the administrative slots of the Washington bureaucracy? In a state legislature? On a city planning commission? Certainly not. In a university? Perhaps. In the employ of an oil company or an independent oil-service-oriented geophysical company? Probably! The best petroleum geologists and related experts are actively wooed and handsomely paid by the oil companies, while their academic counterparts who remain at the university are sought out as consultants and similarly rewarded. The qualified man with no strings attached is exceedingly rare. How else could it be? Far from selling out to the oil industry, the government could ill afford *not* to heed the advice of the oil men.

Representative Richard Ottinger of New York pointed out that throughout government, each agency tends to become the captive of the industry it has to regulate. This is due, in a large part, to the experts also being captives of the industry. Secretary Hickel was sensitive to the fact that he depended on the oil companies for much of his scientific and technical information, and that there might be dangers inherent in this situation. He requested increases in his department's staff and budget to enable the government to come to independent conclusions about oil operations. But the federal government cannot maintain the staff to gather, analyze, and interpret data in *all* the areas where it has a supervisory role. In this sense, each agency must continue to depend on the expertise of its industrial ward.

Unfortunately, this situation has almost always prevailed, and in the case of oil, the industry can hardly be said to have held the public and national interest foremost in its heart at all times. The critical point, of course, is that society's most difficult problems today are not technical ones, but social and cultural and philosophical ones. The dangerous tendency in the past days of science worship was to let the technical people step out of bounds—even to ask them to step out of bounds—and advise us on all matters of how we should live.

The First Advisory Panel

The question of conflict of interest arose in the selection of the panel of experts requested by Hickel through President Nixon. The timing of the request left Science Advisor DuBridge little more than a week to assemble the panel and to arrange for the meeting, although his staff had done some preliminary work in anticipation of the assignment. John Steinhart, a geophysicist from O.S.T., and David Adams of the Marine Science Council were designated as staff assistants to the panel.

Conflicts of interest were more imagined than real. Nevertheless, there were some controversial figures on the panel—men whose presence suggested that the government was once more favoring the industry's point of view. John Calhoun, chairman of the panel, was one of these men. A well-known petroleum geologist, he was Vice-President of Texas A & M University and had been science advisor to Secretary of the Interior Udall. Calhoun was quick to point out that he had departed the Washington scene before the debate over the channel leases began, and therefore had no reason for wanting to justify the subsequent events.

Other panelists whose selection aroused some suspicion were Pan American Petroleum Corporation's Director of Research, William Walton, and Carl Savit of the Western Geophysical Company. In this organization, which was primarily a service group for oil companies, Savit had had much direct experience with oil exploration. The remaining panelists included university biologists, engineers, geophysicists, and oceanographers; and representatives from the Audubon Society, the Haskins Laboratories, the Navy Department's Special Projects Office, and the Dow Chemical Company. (The panelists had been prepared for pickets and demonstrations of one kind or another, public feeling in Santa Barbara being what it was; but the only demonstration to materialize against them was one directed at the Dow Chemical Company because a Dow scientist was on the panel.) All of these men were experts in the sense already described. If the presence of oil men vexed oil's enemies, the choices of a Sierra Club member and a leader in the Audubon Society did just as little for the oil men's peace of mind.

Just what did the press and the public expect from the panel's deliberations? Again and again, Calhoun reiterated DuBridge's charge to the panel: to provide advice and guidelines "for those technological and scientific factors which should be considered to assure safe exploitation of our seabed resources." He patiently explained that the panel had neither been asked nor was qualified to sit in judgment. Assignment of guilt and responsibility were matters for the courts. In fact, the panel could make matters worse by treading in legal territory, when numerous cases had already been filed in the courts and more were certainly to come. The public, however, seemed already to know the answer: Get oil out, get oil out. As the Red Queen put it, "Sentence first, verdict afterwards."

Calhoun did not help when he said that Santa Barbara had been an unwitting and unwilling guinea pig, but that it was important to capitalize on the tragedy for the sake of another time and another part of the country. The "guinea pig" phrase was snatched up by the press and used to heat up a citizenry that was already near the combustion point. Supervisor Clyde was right: He had maintained all along that Santa Barbara was sacrificed for the rest of the country!

Another matter of concern to the press was the privacy of the meetings. Secrecy, they called it. All morning on Wednesday, February 19, the panel met "behind closed doors" in a conference room at the University of California, Santa Barbara. Why was the press excluded? Why were local officials excluded? Why were the Santa Barbara citizens, who had perhaps the largest stake in the affair, excluded? Calhoun tried to describe a work session at which scientists strive, sometimes with a physical and mental and emotional strain amounting to anguish, to get an idea of what to do. They must interact closely with each other, bat their ideas back and forth. They must be free to suggest outrageous solutions, and then consider whether they are really so outrageous after all. It is not a public thing; it is an intensely private thing, and could not occur with an audience. It is how, in the end, the most crucial decisions in government are always made.

In the work sessions, questions were rapid and to the point. Exactly how was the oil coming out of the ocean? No one knew. The two men who had gone down in a submarine for an exploratory look had found the visibility on the sea floor to be zero. Was the leak actually from a fault? Apparently. The bubbles seemed to be rising along a straight line. What was Union Oil's contingency plan? Vice-President John Fraser of the Union Oil Company outlined the strategy, in steps that grew progressively more drastic.

Of Lieutenant Brown, the panel wanted to know the details of legal authority for action and actual chains of events and decision-making. This was not so much to place blame for inefficiency as to discover flaws in the National Contingency Plan and the means for its implementation. Discussion shifted to the chemicals employed, to what was and was not known about their action and toxicity and ultimate fate in the sea.

Then the subject arose of policy on proprietary information related to resource development, particularly in areas where other resources may be in jeopardy. Geophysical information is jealously guarded by the oil companies, who have obtained it at great expense. (It costs in the neighborhood of $15,000 a day to operate an exploration seismic crew at sea.) Oil companies do not part with this hard-won information happily; rather, they sell it to one another. This sort of information, due to the political power of oil, is explicitly excepted in the Moss Public Information Act and guarantee of its proprietary nature is written into federal lease agreements. The most complete and reliable factual and interpretative data, therefore, are inaccessible to regulatory agencies. How and where, then, are the regulatory agencies to find the information they need for developing sound policies?

After two days of intensive meetings an interim report was submitted to DuBridge, and through him to Secretary Hickel. Practical suggestions included procedures for booming harbors; for cutting the kelp beds that were trapping and holding oil, only to release it later and cause renewed contamination of waters and beaches previously cleared; and for disposing of the oil-soaked straw that was used to clean the beaches. There were recommendations for immediate research and development programs in areas where solutions seemed to be technically within reach and where the ephemeral nature of the current oil spill made immediate action necessary if anything at all was to be gained from the sad experience.

Other Activity in D.C.

It is ironical that, at the time of the blowout, the National Science Foundation's Sea Grant Program was considering a proposal from the University of California, Santa Barbara, for a grant to finance a multifaceted investigation of "The Effective Uses of the Marine Resources of the Santa Barbara Area." Its scope embraced petroleum pollution and public policy, geology, economics, and basic ecology. It warned of the possibility of "a catastrophic event similar to the grounding of the Torrey Canyon." After the oil spill, the grant was awarded for an amount and a period of time substantially less than had been requested.

On February 17, Secretary Hickel, anticipating the advice he would receive, granted permission to Union Oil to bleed off gas from its drilling operations on Platform A. Then, on Feburary 20, he startled and delighted the nation by calling a temporary halt to offshore oil lease sales. The only imminent sale affected was the sale of 27 Gulf Coast tracts scheduled for February 25, a sale that would have enriched the federal coffers by millions of dollars.

In Washington, New Jersey's Senator Clifford Case expressed shock on learning that there had never been a public hearing on an offshore oil drilling lease request. He urged Hickel to hold public hearings before granting any more offshore leases. It had been the aim of Congress, he said, to protect the public's interest in offshore lands when it passed the Submerged Lands Act in 1953. Clearly, the public's wishes were not being duly considered. The New Jersey Republican was not merely championing the citizens of Santa Barbara and their fellow sufferers along the oil-polluted Gulf Coast; he felt threatened by news of exploratory activity along the entire eastern seaboard from North Carolina to Maine. He pictured the entire Atlantic coast disfigured by oil rigs and blackened by spilled oil.

On February 21, Hickel, acting on the advice of his panel, authorized the Union Oil Company to put all five of its completed wells on Platform A into production. The rationale behind this move was that if you have a container that is leaking, there are two things you can do about it: you can plug the leak, or you can remove whatever is in the container. There had been numerous futile attempts to plug the leak, and there would be more. The other approach had not been tried, and it seemed, on the whole, a sensible course to take. Union Oil Company's

President Fred Hartley expressed pleasure that, at last, the government had seen fit to recommend the solution that Union had favored all along.

"Everyone Gets into the Act"

One thing was certain. The federal, state, and local governments, the university, citizens' groups, the press, and the oil companies all sincerely wished to stop the leak, to clean up the mess, and to prevent future messes. Anyone who was in any position to do anything was bombarded by letters, telegrams, and telephone calls from concerned individuals who had suggestions to offer. One official reported over 3000 suggestions, of which perhaps 100 had some practical value. Robert Sollen, oil spill reporter for the Santa Barbara *News-Press,* described it this way:[1]

> Almost everybody has a better way to cope with the present Santa Barbara Channel oil pollution crisis than the methods so far employed.
>
> Most of them have offered their advice. Many of them have complained because it was rejected. And some have threatened to sue the Union Oil Co. for stealing their ideas without due credit or compensation. . . .
>
> The News-Press has received its share of suggestions, from an elaborate diagram of complicated devices for trapping the escaping oil to a call from Seattle by a man who wanted to know:
>
> "Have they tried sawdust?"
>
> "Wouldn't sawdust just sink the oil?" the reporter asked.
>
> "Sure," was the reply.
>
> "Wouldn't you just be exchanging pollution on the surface for pollution on the bottom?" the caller was asked.
>
> Yes, but he thought that was an acceptable alternative.
>
> "Why didn't they boom it from the start?" others wanted to know.
>
> The sea was too rough to make booms effective, was the official reply.
>
> "But there are booms with skirts that make booming effective in rough water," the outside experts insisted.
>
> Maybe so, but too little and too late.
>
> Somebody suggested a giant bottoms-up funnel over the seep area, to channel the oil up through the narrow end where it could be pumped out. Somebody else laughed, but that's what Union Oil did. Now some of those who proposed it are incensed because the company stole their idea without acknowledgment, permission, or compensation.
>
> It isn't working too well, anyway.
>
> "Operation Sea Sweep," the big booms forming a giant V which were towed open end forward to trap the oil and pump it up, was a good idea in theory, some experts say, and it would have worked if the company had followed their advice. But Union Oil wouldn't listen. And Sea Sweep didn't sweep up much oil.
>
> Chemical companies across the land came to Santa Barbara with the answer in a steel drum. Their product would dissipate the oil without polluting the waters, they said. Chemical executives issued press releases, called news conferences and conducted demonstrations.
>
> But none could spread their wares on the waters without permission from the Federal Water Pollution Control Administration, and the FWPCA

> was using only the chemicals with which it had had some previous knowledge and experience. It was not in a mood in mid-crisis to start testing new mixtures. . . .
>
> One technique especially criticized by waterfront observers is the washing of the rocky banks around the Santa Barbara Harbor. First, steam-blasting was used. This, it was said, killed much of the bacteria and other organic material that eventually would have decomposed the oil.
>
> Then hot water was blasted on the oily rocks. This spewed fine oil into the air so that near-by boats got a fine coat of oil all over instead of just at the waterline. Moreover, when the tide comes back in, it coats the rocky banks with the same oil that had just been blasted off. So the process starts all over again at the next low tide. It provides employment.

The discouraging thing about the cleanup was just this: The process started all over again at each low tide, or as each change in the wind brought in new oil or redistributed the old. Elsewhere in the country, the oil spill stories slipped into the back pages of the newspapers and then were forgotten entirely. But in Santa Barbara, confronted by the unmitigated reality of the oil, the people did not forget.

VI *Another Blowout*

The very deep did rot: O Christ!
That ever this should be!
Yea, slimy things did crawl with legs
Upon the slimy sea.

About, about, in reel and rout
The death-fires danced at night:
The water, like a witch's oils
Burnt green, and blue, and white.

Samuel Taylor Coleridge

On February 24, 1969, Union Oil Company's well number A-41 on the same platform blew out of control in the Santa Barbara Channel. Once more, oil and gas boiled to the sea surface at an alarming rate. It happened while the well was being reopened after Hickel gave permission to produce the completed wells in order to reverse the pressure gradient and thus slow or stop the continuing leak. Three producing wells on Platform A had been sealed by mud only. Two of these had already been reopened and were producing from shallow sands 400 to 500 feet below the sea floor. Work had just begun on the third well. Its walls had been perforated and the drilling crew was pumping cement through the perforations as an additional safety precaution to assure that the casing was anchored firmly. Suddenly the pressure in the well began to decrease. As quickly as possible, the crew pumped 100 sacks of cement into the well. The pressure continued to decrease. At the same time, the seep at the northeast corner of the platform began to bubble violently. Two hundred more sacks of cement were pumped in, then

1300 cubic feet of mud. The pressure kept dropping and the bubbling grew worse. Perhaps the end of the casing, instead of being set in solid rock, was situated in an oil-bearing stratum. Or perhaps the drilling crew, in squeezing cement through the perforated casing, had exceeded the lithostatic pressure and had forced open a new channel to the sea floor. In any case, oil began to escape rapidly from its prison.

Although the core of the new disturbance was about 100 feet across, the surface of the water was broken by dark boils of blue-green oil over an area 500 feet in diameter. The oil engulfed the platform itself and extended to the west in a dense river. By February 26, the river of oil measured 8.5 miles long by as much as 1000 feet wide.

Union officials were in no hurry to spread the news of the new spill. Lieutenant Brown heard rumors of an increase in the leak, but he was unable to confirm them because violent storms had again pummeled the area, forcing cancellation of several Coast Guard surveillance flights. Heavy concentrations of oil were once more reaching the beaches near Santa Barbara, although much of this was old oil that had been broken loose by the storm from its entrapment in the kelp beds. Beach crews were overwhelmed by the pitiless advance of the black tide. By February 26, Union officials had to admit that well A-41 was probably permitting deep zone oil to enter into the formation located about 900 feet below the bottom of the sea, and subsequently, through some fracture or fissure, into the channel floor.

Preliminary attempts to seal well A-41 failed. There was then a delay of two days while new supplies of mud and mud barges were sent from Los Angeles. Vestiges of oil from the early days of the spill joined with newly released oil in the worst assault yet on the beaches. Access to several beaches was prevented by flooded roads and storm debris. No one knew how to clean up the mess from the oil and the storm. First one thing, then another, was tried. A six-foot high sand barrier was built along the Goleta beach as an experimental block to the oil. Some observers commented that the wall might be wiped out as easily as sugar. But again, it might not. Someone rigged up a grader with a rake-like blade, to rake the oil-covered sand and debris into the surf. In this way the garbage was separated from the sand and everything got a good washing. Wave action then swept the debris back onto the beach, where it could be collected and disposed of. On another beach a caterpillar pulled a truck which, in turn, towed a straw spreader. Two men fed straw into a cannon-like chamber while a third operated the "gun" to spew the shredded straw onto the oiled sand. More workers followed to rake it up again. It lay in ugly piles where it was collected, serving as a grim reminder of the oil spill and a considerable fire hazard, as well. Increasingly often, after the hassle over what to do with it was settled, it was collected into piles on the beaches and burned. Storms and floods continued to add to the general chaos. The Ventura marina was washed out by a flood which contributed to the junk in the channel, among other things, one dead cow and one dead bear.

Until mid-March, the leak first increased, then decreased, without any apparent cause and irrespective of what was or was not done to control it. Well

A-41 was finally plugged on March 1 and the leak decreased for a time to its baseline rate of (according to Union's possibly conservative estimate) 2000 to 4000 gallons per day.

On March 2, Union confessed that attempts to reactivate well A-21 had set off a new leak, but that the well was again plugged, this time, upward from the 3000 foot level. The next day, however, the oil flow mysteriously increased to 10,000 gallons per day.

In his Sitrep 52, Lieutenant Brown pessimistically noted that since "no significant progress has been made in stopping or containing the seepage since 7 Feb. either from the platform or by known containment methods (i.e. booms, chemicals, skimmers); and . . . the Union Oil Co. solution of drilling wells may take from one to two months to show any effect; and . . . the caisson/rock and cement blanket would take one to two months to construct; it was decided to adjust operations for a minimum of two months and possibly longer."

Other Acts in the Show

While the struggle to stop the leak and clean the beaches continued to occupy center stage, other minor dramas were being played out in the wings. Almost simultaneously, 15 state tidelands exploratory oil permits in the Santa Barbara Channel were revoked, and citizens in Orange County to the south learned that exploratory drilling off their coast was about to begin. Reportedly, the state had quietly issued the permits to Shell Oil before the local citizens were aware that drilling was being considered. It is only a geological survey, they were told, to quell their anxiety. Orange County citizens had followed the plight of Santa Barbara closely, wondering how long it would be until they suffered a similar fate. "A geological survey!" one citizen commented wryly. "It always begins with a geological survey."

On March 3, the Interior Department's Eugene Standley announced that 27 wells on Platform Hogan off Carpinteria were being put back into production to relieve a dangerous pressure buildup.

The following weekend, GOO (Get Oil Out) sponsored the bitterly satirical melodrama "Barbara's Fatal Fault," starring Sir Derrick, villainous Earl of Union, Uncle Sam and his hired hands Fickle and Stu, a lion-hearted newspaperman named Scoop, and a host of other characters whose resemblance to actual persons was devastatingly intentional.

On March 6, 1969, the first oil from the Santa Barbara oil spill reached San Diego.

President Nixon visited Santa Barbara on March 21. A few picketers carried signs reading "Welcome to Santa Barbara's Dead Sea: a project of your Federal Government." After touring the stricken area by air, Nixon made a sympathetic and conciliatory public statement. It was the kind of statement which impresses you as you first hear it, but which you discover, on rereading it, really said nothing at all.

The same day, Walter Hickel issued a statement about the oil spill and what he had done in response to it. The statement was applauded by both his friends and critics, liberals and conservatives:

> Today, almost eight weeks after the oil well blowout in Santa Barbara Channel, I am certain we all agree that the pollution of California's coast, without a doubt, was a tragedy. . . . And all of this may have been unnecessary.
>
> Yesterdays always are easier to examine than tomorrows. Officials who a year ago allowed oil production in Santa Barbara Channel would have welcomed a preview of recent events. Given a look at the tomorrow, the economic pressures which dictated the $603-million lease sale at the expense of our environment would not have prevailed.
>
> But that was yesterday. The mistakes of the past can and will serve as the basis for a better tomorrow.
>
> There has been a whole new approach to the nation's offshore oil situation during these past eight weeks:
>
> On January 24, I was sworn in as Secretary of the Interior. Four days later the oil "blowout" occurred.
>
> On February 2, I flew to Santa Barbara for a first-hand look at the problem, and on February 7, with concurrence by the President and the Department of Justice, I ordered all wells closed and drilling halted in those Federal waters.
>
> I then initiated a broad review of all Outer Continental Shelf regulations. . . . Oil companies were invited to look at our proposals to see if the industry itself could suggest ways of tightening the regulations–and they did propose some tougher rules.
>
> Governor Reagan agreed to have his oil and gas officials provide an additional review of our new regulations. This review . . . served as a first step toward closer voluntary coordination between State and Federal agencies concerned with the common objectives of preventing pollution. . . .
>
> I ordered a moratorium on all scheduled oil leasing on the Outer Continental Shelf, and we are reviewing the criteria upon which decisions to lease offshore lands are made.
>
> I issued an order making oil companies absolutely liable for pollution if it occurs in the future, regardless of whether or not they are at fault.
>
> At my recommendation, a panel of distinguished scientists and engineers was named by the President to make recommendations on cleaning up the pollution in Santa Barbara and on fighting possible future pollution. Many of their ideas already have been followed. . . .
>
> And we are pushing for new legislation to strengthen the Federal Government's role with regard to oil pollution and spills.
>
> With our modern technology, we can get at the source of the problem. First, we must discipline our procedures, and in the form of OCS Order No. 10, I am putting into effect today the most stringent possible regulations we can devise to safeguard our environment. . . .
>
> The second major action I am taking today is the signing of an order which turns the existing two-mile buffer opposite the Santa Barbara State Oil Sanctuary into a permanent ecological preserve. . . . The new Santa Barbara Ecological Preserve is 21,000 acres. . . . This area . . . is extremely valuable for fishing and other recreational uses.

> In addition, all unleased areas south of the Santa Barbara Ecological Preserve will be held as an additional buffer zone. No drilling or production will be permitted in this 34,000 acres. . . .
>
> About half of the remaining Federal lands in the channel are not leased. Before any consideration is given to leasing these areas, the public will be consulted and its recommendations carefully considered.
>
> . . . Two weeks ago, I requested oil companies with leases in Santa Barbara Channel to furnish the Department of the Interior with complete geological and engineering information. . . .
>
> . . . After satisfying ourselves, as far as humanly possible, that accidents of the Santa Barbara type might be prevented, we will allow the oil companies to resume operations in their present locations. . . .
>
> . . . Our efforts do not end today. We will continue to look ahead to tomorrow's environment as much as we look at today's development of our resources.

A powerful statement from a man who, in a few months, moved from a position of skepticism to a firm stand as one of the country's leading conservationists. The Santa Barbara incident may have been the greatest challenge of Hickel's aborted term as Secretary of the Interior. When he acted, he acted decisively. Partly because of his penchant for taking forthright stands on issues, Walter Hickel was later fired by President Nixon.

Misfortune continued to dog Union's efforts to contain the leaking oil. By March 5, they had installed a cylinder under the platform to trap the escaping oil as it bubbled from the sea. During the night of the 5th, stormy seas bashed the cylinder against the legs of the platform until it was demolished. At the time, a new, larger steel funnel was being built to supplant the original one. The new hood was towed to the platform but it ruptured a seam during installation, and was taken back to shore for repairs. Oil continued to leak unconfined. Well A-21 was reopened for a third time, this time to the 850-foot level, to reduce pressure in shallow strata and, hopefully, to reduce the leak.

For more than two weeks, divers and workers from the barges Wodeco I and Lindsay tried unsuccessfully first to find and then to cement shut leaking fissures in the sea floor. This operation was continually hampered by low visibility and water turbidity. After ten days there was no indication that this effort was having any effect at all.

On March 18, attempts to repair the damaged hood were abandoned and a smaller hood was installed over the main leak. On the same day, Union resumed application of chemicals around the platform, in accordance with a new set of F.W.P.C.A. regulations established specifically for the Santa Barbara pollution incident.

Lieutenant Brown and other officials feared that Union's public relations work was lulling Santa Barbarans into thinking that things were improving and would soon be back to normal. Warm weather and an influx of spring vacationers were certain to cause increased public concern and protest. Still, there was nothing more that the army of government servants could do. On-scene forces met in Santa

Barbara on March 24 and recommended that the incident be suspended on March 31 and the R.O.T. deactivated. There was an air of defeat about the decision to pack up and go home. They had done what they could. It was not enough, but there was nothing more to be done.

On April 1, the Santa Barbara County Supervisors declined Secretary Hickel's invitation to comment on his new offshore regulations. They voted instead to inform Hickel, Nixon, and other officials that their position had not changed, that they wanted oil *out.* The Supervisors feared a trap. They feared that approving the regulations would be seen as tantamount to approving drilling, and they wanted no more of that.

On April 2, Hickel lifted the ban, effective since February 7, on drilling on five channel leases. The shutdown order was still in effect for 67 other leases. Still, it seemed only a matter of time until activity would be resumed on all the leases. County Counsel George Kading hastily put the finishing touches on a petition for a mandatory injunction to halt drilling in the channel. He had been working on the papers for weeks, in anticipation of this event.

Within minutes of the official OK signal, two drilling ships stationed in the channel began operations. Bluewater II resumed work on the hole she had been drilling at the time of the halt. Wodeco IV began a new hole. The next day, 23 wells on Platform Hogan were put into production again. And the day after that, April 4, George Kading carried his 13-page injunction to Los Angeles, to file in the United States District Court.

Who Goofed in Santa Barbara

A series of errors and unlucky coincidences caused the Santa Barbara blowout. The very choice of the drilling team was unfortunate. Most oil companies do not do their own drilling; they contract the job to an outside drilling company. In the case of Santa Barbara, the Union Oil Company sold the job to the Peter Bawden Drilling Company of Canada, for the reason that Peter Bawden submitted the lowest bid. Perhaps they bid low because they did not realize what was involved in drilling in the channel, for they were generally unfamiliar with the area and had never before drilled off the coast of California. This was a matter of some concern to Regional Supervisor Solanas, but in the end he allowed Union to go ahead with its plan.

Then an exception was requested, and granted, to omit the installation of intermediate casing. The problem in Santa Barbara was too much oil, oil too close to the surface. There was no reliable place to anchor the intermediate casing. Under conditions like that, in the event of an accident in the area of porous sandstone, loose casing could be a greater hazard than no casing at all. Hence, the application for "variance" was granted. This was not unusual. Exceptions are frequently made when a well presents some unique problem. It would be surprising, in fact, if any blanket set of regulations could be designed to apply to all conditions that might arise. It was revealed during later investigations that applications for variance are rarely, if ever, turned down by either federal or state authorities.

So well A-21 was to be drilled without benefit of protection from intermediate casing. This in itself was no cause for alarm. It was the fifth well to be drilled from Platform A under these circumstances. The first four wells had created no problems.

Like almost half of the blowouts that occur, this one began as the drill pipe was being withdrawn from the hole. In all probability the driller was removing the pipe too rapidly—a tendency of many drillers when operating at relatively shallow depths such as this. Swabbing the hole would cause high pressure gas to be drawn into the well bore and to permeate the drilling mud, further reducing the pressure in the hole. The crew was having trouble maintaining pressure; the mud was dropping low in the hole. They should have stopped right there and filled up the hole, before withdrawing another foot of pipe. By that part of the operation they had run out of drilling mud of the proper density and had no way to adjust the density of the material they had on hand, so when things began to go wrong they threw a lot of low density mud into the hole and hoped for the best.

When the blowout began, the crew was unable to remove a section of drill pipe that was stuck in the derrick. They were unable to screw in a 75-pound safety valve. They broke part of the kelly valve assembly, eliminating the possibility of control through the kelly. They finally were able to drop the drill pipe back into the hole and close the blind rams on the blowout preventer, sealing off the well. But the well was then completely sealed off, and this was the most disastrous error of all. The crew had violated a cardinal rule of drilling: never completely shut in a well in trouble. The worst thing to do with a blowout is to shut it in. The instant the pipe rams were closed on the pipe, a choke should have been opened to allow the well to vent from the casing annulus. Permitting limited venting under careful supervision prevents excess buildup of back pressure against the formations down hole.

Remember that well A-21 was largely uncased. When the well was closed and excess pressure was prevented from escaping through the wellhead, pressure mounted in the well bore below the casing until the pore fluid pressure in the strata below the casing was exceeded. Pressurized gas and oil began to force their way from deeper sands into the shallow permeable beds. Now the pressure of the newly energized pore fluids in the shallow sands built up to the hydraulic fracturing pressure of the caprock. The caprock ruptured and the sea began to boil with escaping gas and oil. The inappropriate remedial actions of the drillers had led to violent rupturing and fracturing of the earth's crust. A similar phenomenon probably occurred in the blowout of well A-41, when the careless pumping of cement led to disaster.

Thane McCulloh, staff geologist for the U.S.G.S., described the appearance of the surface and floor of the sea, as he observed them on a calm day in late April 1969:[1]

> The otherwise smooth water surface was ruffled over a broad area by myriads of small breaking bubbles of gas accompanied by rising globules of light oil that quickly spread as they surfaced to form iridescent films streaked with

> brown. On the sea floor the oil issued as streamers from pinpoint and larger openings in siltstone outcrops and soft sediment alike. Pits and elongate craters on the sea floor ranged from a few feet to many feet in length and up to several feet in depth. In this vicinity the sea floor was littered with angular rock debris, . . . presumably ejected forcefully from the sea-floor depressions at the time of the blowout eruptions.

This accident could and should have been prevented at any one of several steps along the way. But as chance after chance passed by, the fate of Santa Barbara was finally sealed with the sealing of the well, less than fifteen minutes after the kick began. Many people clung to the belief that the blowout never would have happened if the federal regulations had been as stringent as those of California. This is not true. Comparison of the two reveals few differences. Two *similarities* between state and federal practices may have played a more significant role than differences did. First was the practice of granting variance requests almost routinely. The rules make very little difference, if exceptions are freely made. Second was the relatively less rigorous inspection of offshore operations than of onshore wells. This is a widely recognized problem. Inspection of onshore procedures and equipment is usually thorough because it is easy. Not so, offshore. The *Los Angeles Times* reported that there had been no federal inspection of well A-21 before the blowout.

The Advisory Panel Advises

While the oil leak in Santa Barbara dragged on into its second, third, and finally fourth months, the President's Advisory Panel met two more times to lay strategies for the long-range war against oil pollution. At their first meeting they had given their best advice about what to do in Santa Barbara; now they were considering the remaining aspects of their assignment. They met on March 4 and 5 in Los Angeles and again on April 15 and 16 in Washington, D.C. They asked searching questions of each other and of their consultants from government, industry, and academia. Some of these questions remain unanswered. The probability of an oil spill can never be reduced to zero. What, then, is an acceptable risk? What are the relative risks from, and the state of technology of, offshore drilling, pipelines, tankers and oil barges, and onshore facilities? How can tanker design and operation be improved to cut down risks? What constitutes acceptable cleanup after a spill? What will be the impact of decreased offshore production and increased cost of production on the nation's economy? Since chronic low-level pollution is, in the long run, a greater ecological threat than an infrequent spectacular spill, what standards for water quality are attainable—and how? When will the Coast Guard and the F.W.P.C.A. have a workable set of effective operating procedures and contingency plans in standby readiness?

The panel's two-part report was issued in the fall of 1969. Part I provided a set of recommendations for reducing the probability of an oil spill and for cleaning one up should it occur. The pervasive mood of the report, which amounted to a plea, was, let us begin at once to do the research that will enable us to clean up an oil spill. This included development of techniques for monitoring a spill and predicting its movements, for physical and chemical containment of a spill, for salvage of spilled (or, in the case of a tanker, potentially spilled) oil, and for beach cleanup and restoration. Fundamental to all of these needed technologies was the mandate to minimize ecological destruction.

Part II of the report, entitled "Offshore Mineral Resources: A Challenge and an Opportunity," described a philosophy for the development and conservation of natural resources. It proposed an intermediate course between "exploit now, without heed for the future," and "hands off, for all eternity." The recommendation was for the recognition of a class of escrow resources—resources whose development would be postponed for a fixed period of time, after which the situation would be reviewed. The result of the review could be either to proceed with development or to renew the period of reprieve. Basic to this idea was the realization that both technology and human needs change. What today is an ugly oil rig might evolve in five years to an almost foolproof sea-floor installation, eliminating present objections to oil production. What today is a sparkling beach inhabited by a myriad of marine creatures and visited by crowds of happy bathers may be, ten years from now, an industrial cesspool—so what matter if the oil industry then moves in? And in the more distant future, 20, 30, or 50 years from now, the ready availability of almost unlimited power from fusion might make the meteoric story of oil seem as quaint as the saga of the horse-drawn streetcar. Several bills introduced to the Congress since the oil spill have incorporated the idea of escrow resources. Notable among these are legislation proposed by Senator Gaylord Nelson of Wisconsin and by Rogers Morton, Walter Hickel's successor as Secretary of the Interior.

And a Second Panel Is Organized

But the oil continued to leak, while the panel occupied itself with panoramic thoughts for the future. The oil continued to leak, while half a dozen major oil companies wondered about the future of their Santa Barbara Channel leases. The oil continued to leak, and Santa Barbarans seemed to overlook this fact in their emotion-driven campaign to get oil out.

On March 27, acting for the Secretary of the Interior, Russell Train requested the President to seek the advice of a group of independent geologists and engineers on the future of the Union Oil Company lease. In short, the problem of the leak had not been solved, the future of the Union Oil lease had not been determined, and the Interior Department wanted the best brains in the country to advise them and to back them up in their decision on what to do. The result was that, on April 7, 1969, President Nixon (through his advisor John Ehrlichman) requested Science Advisor DuBridge to assemble a second panel of experts. Once more letters and telephone calls went out to eminent scientists and to Fred Hartley, President of Union Oil. The Union Oil Company was asked to present its plan or plans for stopping the persistent leakage in the channel and for insuring against future leakage, as well as its plan for withdrawing from the lease in the event the leakage could not be stopped. John Calhoun, who had chaired the first panel, was invited to chair the second also. Other panelists were selected and carefully quizzed about their oil company stock holdings, consultant jobs, and other oil-related work. The panel meeting was set for May 12 and 13, 1969. The immediate future of Santa Barbara hung in the balance.

VII A Controversial Report and a Diminishing Leak

It's the same old story
Everywhere I go,
I get slandered,
Libeled,
I hear words I never heard
In the Bible.
And I'm so tired,
I'm oh so tired,
But I'm trying to keep my customers satisfied,
Satisfied.

Paul Simon

During the spring of 1969, there was so much political activity, so much passionate energy, devoted to banishing oil forever from the Santa Barbara Channel that the problem of how to stop the persistent oil leak went all but unnoted. Oh yes, people were aware that fresh globs and streamers of oil continued to ooze from the sea floor, but this awareness served merely to inflame their wrath and further stimulate their anti-oil efforts. Most people were unwilling or unable to confront the problem rationally. Oil was leaking; how could the leak be stopped? Psychologists have a name for this kind of inappropriate response. They call it displacement activity. You can't confront the real problem so you kick your dog or bite your nails, instead.

The Second Panel and Its Report

The charge to the second presidential panel on the future of the Union Oil lease was simple and profoundly clear: considering only the technical aspects of the situation, present and future, make recommendations for stopping and permanently eliminating the seepage in the channel. In a letter to John Steinhart of O.S.T., William Ruckelshaus, then Assistant Attorney General, warned, "It must be made absolutely clear to the panel that they are not to conduct a 'hearing' concerning past events; they are to direct their attention solely to future operation of the Union Lease." Thus, the panel was explicitly prohibited from assessing blame for the original blowout. By then the question of blame was central to lawsuits and counter lawsuits with total damages claimed amounting to billions of dollars. Furthermore, the panel's attention was directed to the leaking Union Oil lease *only*, not to the other 70 leases in the channel. To make certain that the panel did not stray from its narrowly defined task, the Department of Justice sent two representatives to attend the meeting, to be on the alert for prejudicial statements by the panel.

It is worth looking straight at the predicament our society and our government face in this case. Our adversary legal system encourages suits for damages real or imaginary. In this way we honor our search for nonviolent redress of grievances. But this well-intentioned device has been corrupted through the hyperbole of charge and countercharge and the demand for inflated damage payments. The rules of evidence, adversary practice, and the quest for justice uninfluenced by information outside the courtroom all urge the litigants to forego publicity and to keep evidence to themselves, to avoid compromising their case. In the same letter to Steinhart, Ruckelshaus said ". . . the publicity and information releases from the panel must be strictly controlled. An adverse report from a blue ribbon panel such as this could be very detrimental in the pending and anticipated litigations."

We accept such tactics with relish, in Perry Mason stories. But when one of the litigants is the federal government, these practices appear secret, closed, manipulative—or worse. Preventing prejudicial information from getting out may require news management, news blackout, or manipulated stories. The trouble lies in the system; and it is a true dilemma, for the alternative is no better. A government that gives its own case away by publishing evidence and tactics when the other side does not, and that permits its employees to prejudice its case (whether for well-intentioned or for nefarious reasons)—such a government is both easy prey for the unprincipled and vicious fragment of its populace, and unresponsive and unfair to its honorable citizens.

The second panel met in Los Angeles, on May 12 and 13, 1969. In preparation for the meeting, the Union Oil Company had sent each panelist a folder containing its proprietary technical and geological information on the Dos Cuadras field. At the meeting, Union outlined its plans for stopping the leak. By now, the oil company was desperately committed to finding a solution to the problem. It was the only way out from the constant pressure of the government and the news media.

In two days of deliberations, the panel considered many suggestions for controlling the oil. An assistant professor from the University of California at Santa

Barbara wanted to pave about 20 acres of the ocean floor. That did not seem very practical. A slab of concrete (or anything else) of that size would not be durable anywhere, but especially not in an earthquake-prone area. An Atomic Energy Commission employee suggested setting off a 20 kiloton bomb in the channel, to create a reservoir for the oil. Imagine Santa Barbara's reaction to that idea, had it been publicized! First they drown us in oil, then they bomb us! The Army Corps of Engineers thought that with $25 million they could build an artificial island over the area of the leak. It would take them quite a long time—time reckoned in years. They were not sure what would become of the oil even after the island was built.

The problem was still the same as before. There was a container, weak and fractured, and it was leaking its contents. For three and a half months, unrewarded efforts had been made to repair the container. It looked as though the leak could go on indefinitely, for centuries, perhaps. The remaining alternative to plugging the leak was to empty the container as quickly and as completely as possible. In the end, that is exactly what the panel recommended.

Other temporary measures were also suggested, to be undertaken along with the systematic drilling and production of the entire Repetto formation. Earthquake hazards were considered and the seismologists, structural engineers, and geologists concluded that the issue had been blown out of proportion and constituted at most a minor risk. Nevertheless, the public had no way of knowing this, and was probably not ready to believe it; so the panel urged attention to earthquake hazards.

The panel was not permitted to say that the carelessness of the drilling crew had created the mess by actually fracturing the rocks that had been restraining the oil, because they were not supposed to concern themselves with past events. So they contented themselves with suggesting that the entire pumping and grouting program be carried out under close supervision of the Department of the Interior.

Details of production remained to be decided. The safest course seemed to be to produce all the Repetto reservoirs at one time, throughout all the levels that were contributing to the leak. The decision on how deep to go depended on which model for the structure of the area was selected, but the oil company and U.S.G.S. scientists agreed that there was one particular model, from among the possible ones, that best explained the available data. The scientists also agreed that draining the entire oil field at once would prevent the development of potentially hazardous pressure differentials between formations. It was decided to replace the oil that was drained from the reservoirs with water. This would have the multiple advantages of minimizing the subsidence of the earth where oil was withdrawn, of increasing the total percentage of oil recovered, and, by saturating the porous structures with water, of reducing the relative permeability to oil to a point where no free-flowing oil phase could exist. Any future seeps would then be water, not oil. Production was to continue until all available hydrocarbons had been removed. It was felt that if the wells were abandoned as soon as the leaking stopped, it would be but a matter of time until pressure would again increase and seepage would resume along the old, still-existing fractures.

The Public Outcry, and Some Cooler Minds

The panel was unanimous in its endorsement of further drilling and production on the lease. As one member put it, he would sleep better nights if he knew that pressure was being withdrawn from the field and he did not feel that it was a bomb ready to go off.

The bomb, if there was a bomb, was neither the one suggested by the A.E.C. representative nor the one feared by the panelist, but the release of the panel's report. The Santa Barbarans were bitterly disappointed. They had felt sure that everyone would see as clearly as they did that offshore drilling had no place in the channel. Instead, a stepped-up production program was proposed. Worst of all, the Department of the Interior and the Union Oil Company, seen as equally villainous partners in the rape of Santa Barbara, instead of being punished actually stood to gain financially from the continuation of their evil-doing. This unfortunate aspect of the suggested remedy was the most damning of all. From the viewpoint of the average citizen or local official, the situation was diabolically clear. Through their greed and ineptitude, the government and the oil company had caused the trouble in the first place. They then got together and worked out a "solution" which was to be mutually profitable for them both.

The frustration of the press and public was expressed in a wave of editorials that swept the country. Almost everyone knew he did not like the proposed solution, but no one could show that it was wrong. It appears that the technical correctness of the solution was not the real issue. Santa Barbarans wanted an end to oil production in the channel, and they wanted to make the decision locally. Here are a few samples of the journalism that followed the release of the second panel's report.

From the *San Francisco Chronicle,* June 5, 1969:

> While a technical judgment of this sort cannot convincingly be disputed except by other technologists, there are plenty of laymen and public officials in Santa Barbara who don't like it. Oil is still seeping out of the Union platform area at a rate of at least 1500 gallons a day and washing up onto Santa Barbara channel beaches.

From the *Chicago Sun-Times,* June 4, 1969:

> The recommendation that Union Oil Co. be allowed to continue pumping oil from its leaking offshore well at Santa Barbara, Calif., until the oil deposit is exhausted may be technically sound but it inspires no confidence.

"Sacramento Bee," on June 9, 1969, observed:

> No oil company which has been a party to such a major disaster for so much of California's shoreline should be rewarded, thus, by being permitted to continue drilling on an even larger scale. To say this is one way of

eliminating the oil seepage—to remove all the oil from the underwater structure—begs the main questions involved. These are: What caused the mistake and how can it be avoided in the future?

From the *Detroit News,* June 4, 1969:

> Perhaps the recommendation by a presidential panel that pumping be resumed in the ocean off Santa Barbara, Calif., from an oil field that sprang a leak last January is geologically correct. But the stakes are so big that the Interior Department should seek additional advice. . . . Once the drilling commences and the leak occurs, the only remedy appears to be to continue drilling. . . . To the logical, though non-geological mind, something doesn't ring true.

Both Secretary Hickel and Science Advisor DuBridge tried to reason with their irate correspondents. In a letter to Louis Lancaster, president of Santa Barbara's Channel City Club, DuBridge wrote:

> I am distressed to learn that the residents of Santa Barbara are upset about the recommendation of our expert panel on the oil leakage. It seemed to me such an obvious and wise suggestion that if there is a large pool of oil under pressure under the Santa Barbara Channel and if the cap over this dome is leaking, the best thing to do is temporarily plug the leaks and rapidly remove the oil. . . . No one has suggested any other solution that I have heard of.

The press also produced some insightful analyses of the situation. As the *San Diego Union* of June 5, 1969, saw it:

> Current events illustrate one of the human problems that persists (sic) as solutions are sought to eliminate the troublesome oil leakage in the Santa Barbara Channel.
>
> A panel of scientists suggests the best way to control the seepage is to authorize resumption of drilling in order to relieve pressures within the oil basin. The immediate reaction was to challenge the findings as the product of inadequate research or to impute bias on the part of the scientists.
>
> Yet Dr. Lee DuBridge, who led the study, and all other persons and agencies concerned with the Santa Barbara problem have acted responsibly since the disaster.
>
> The fact simply is that there has been a substantial communications gap over the Santa Barbara oil pollution. Questions being asked by concerned people are not being answered. . . .
>
> And people find it hard to understand why the technology of the United States cannot devise safeguards to trap escaping oil.

The *Washington Post* of June 7, 1969, pointed out that

> . . . the problem will not be solved by mere emotional reaction against the pumping-out idea because it would enable the company and the Government to profit from the operation. Unless some other feasible solution can be found, there seems to be no acceptable alternative.

Most of the specific criticisms of the panel were expressed by California's Senator Alan Cranston, who was joined by Senator Edmund Muskie of Maine in a lengthy statement to the press. Cranston's behavior was not surprising. He was rising to the defense of his constituency and making a little political hay at the same time. Muskie's actions, however, were probably ill-advised. Fresh from his rapid ascent to fame and the adulation of many young people and political liberals in the 1968 election, the defeated Vice-Presidential candidate and potential Presidential candidate was eager to enhance his reputation as Mr. Environment. To this end he hopped, somewhat impetuously, onto the Santa Barbara bandwagon. He had not done his homework. Through his aides, who had been contacted by Cranston's office, Muskie first heard about the press conference on the day it was held. A few of Cranston and Muskie's objections and the answers to them are summarized below.

"The panel . . . met only once, at a two-day meeting. . . . The recommendations agreed upon were based on drafts which were circulated by mail during the following two weeks." There is nothing *a priori* wrong with that. After considering all the information, the members agreed on the best course of action. Most of the panel members had studied the Santa Barbara problem before the meeting. Five of them were also members of the first panel, which had met in February, March, and April.

". . . millions of barrels (of oil) will probably remain in the well after all the pumping is finished because the industry does not have the technical means to exhaust all the oil from a particular reservoir. The remaining oil could very well continue to rise through the open fissure and continue to smear the beauty of Santa Barbara beaches for still another two or 10 or 20 years."–Cranston and Muskie may be very good legislators, but they are very bad geologists. That is not at all how the residual oil would behave.

"Did the panel make a detailed study of the tectonic make-up of the channel floor to support its conclusion?"–Yes.

"Did the panel make a critical evaluation of alternative methods of stopping the leak?"–Yes.

Still More Experts

The 1968 presidential race divided the Republican party in California. The split between Nixon and Reagan supporters at the time of the convention was officially smoothed over, but a certain tenseness and mistrust persisted. Only by continuous contact at the staff level was an open break prevented over the publicity attaching to the recommendation for further drilling and production on the Union Oil lease.

The California Advisory Commission on Marine and Coastal Resources met three days after the release of the report. Discussion was long and acrimonious. The laymen on the Commission, mostly identified with conservationist views, wanted to reject the report out of hand. They ignored the question of the continuing leak.

The scientists and engineers were disappointed or angry at the lack of detailed background data in the report, but they urged calmer consideration. Many knew members of the panel personally or by reputation, and they were confident of the panel's integrity and ability. The outcome was a compromise. "The merits of the specific recommendation . . . to drill additional wells . . . cannot be judged until the full report of the Panel is available." They called for delay awaiting "adequate expert review" of the full report. Alas, in the face of Moss Act and Department of Justice requirements, there was no full report. The call for experts to review the experts who study the work of experts is yet another case of failure of our system in dealing with problems to the satisfaction of the people. There is no solution to the problem of watching the watchers.

The City Council of Santa Barbara had its own experts. Upon request, the Environmental Quality Advisory Board examined the situation independently. A report prepared by Dr. Philip W. Reinhart was endorsed fully by the advisory board and forwarded to the City Council on August 11, 1970. The report looked in detail at the continuing but diminished leak, including the results of the grouting, drilling, and production program recommended by the original report. To the chagrin of Santa Barbarans, the conclusions endorsed the production program as the only solution to a bad situation; and–horror upon horror–it recommended more drilling and production from yet another new platform to fully deplete the Dos Cuadras field. Drilling elsewhere in the channel was opposed.

More About Secrecy and Public Information

The "credibility gap," as it has been called, is one of the serious problems confronting our society today and undermining its morale. The government never admits to spying until a spy plane is caught. It never acknowledges its research on chemical warfare until thousands of sheep die mysteriously. It is notorious for suppressing reports that are unfavorable to current administration policy.

In recognition of this, the Moss Public Information Act was passed in 1968. Its purpose was to assure the public of access to all facts, interpretative information, and opinion that are available to the government. If it is written down, it is in the public domain, and the popular "for administrative use only" stamp carries no weight. In spite of the act, however, it is possible for the government to hide a great deal simply by being very quiet about the existence of information. We can paraphrase an old adage as Big Brother might see it: What you don't know you don't know won't hurt you.

A few kinds of information were specifically excepted by the Moss Act, and they included the proprietary data of the oil companies. If the loopholes in the Moss Act are abused, there is the possibility that the term "proprietary" will come to be used to camouflage domestic issues in the same way that "classified" and "secret" are used by the military. But at the moment, that is the law.

Like the first panel, the second was widely criticized for its mode of operation. Fred Eissler, National Director of the Sierra Club, was particularly outspoken in his opposition to the closed door meeting and its uncontested, unsupported (by accompanying data) conclusions. The press made a big thing of Union's secret information that neither the oil company nor the government nor the panel would make public.

The ironic thing is that no one really wanted to know the technical issues or to debate the technical questions. People were merely angry because they could not have the information. Secrecy connoted nefarious schemes and collusion between the government and the oil company. Reporters were given a 400-item bibliography on the Santa Barbara Channel, representing publications from the open scientific literature. There was no indication in the resulting press coverage that any of the reporters had made use of this information. Under pressure from those demanding information, William Pecora announced that the U.S.G.S. would present the "facts" in a special professional paper. When the report came out in October 1969, it was hardly a best seller. Some three copies were sold in Santa Barbara and fewer than fifty in all of California, most of them going to the state government. No, Santa Barbarans were not interested in technical reports. They wanted oil out of the channel, but most of all, they wanted to feel that the decision was theirs.

How Much Oil Leaked?

Two gross deficiencies in our technological competence were highlighted by the oil spill. We were unable to predict the movement and behavior of an oil slick, and unable to estimate the amount of a spill. When a tanker carrying 100 million gallons splits open and pours its cargo into the sea the answer is easy: it spilled 100 million gallons. For oil pouring out from the earth, the answer is not so readily available.

During the early days of the oil spill, the Union Oil Company estimated the rate of leakage to be between 15,000 and 20,000 gallons per day. This estimate is almost certainly low. The oil company gave its most optimistic appraisal. Chemicals were applied to the slick in amounts that should have sunk or dispersed it if the quantity of oil was anywhere in the range of Union's official estimate. But the slick could not be brought under control with chemicals.

Other observers set the rate of leakage as high as 400,000 gallons per day. This guess is almost as certainly too high as the first was too low. The true figure is somewhere between 20,000 and 200,000 gallons per day. A leak rate of about 45,000 gallons per day during the most severe periods of the spill falls within the range of two different determinations which yielded the conservative and the extreme estimates. With the exception of a few short periods, this high rate of leakage persisted until mid-March. From then until mid-July, the leak rate averaged 1000 to 1500 gallons per day. During the last two weeks of July, probably as a result of the cement grouting program or the production of the shallow reservoirs or both, the leak rate dropped first to 800, then to 600 gallons per day. By early September 1969, the seep was reduced to less than 400 gallons per day, and it has been declining very slowly since then.

It is difficult to determine the amount of oil in a slick because the thickness of the slick may vary from little more than a monomolecular layer riding the surface of the water to a river of oil inches deep. The thickness of the slick depends on the composition of the oil and on wind and current conditions. Many of the techniques used to map the area covered by the slick also give an indication of the thickness of the oil layer, but more research is needed before any of these methods becomes reliable in a quantitative sense.

Among the techniques that show promise are two types of determination by radar. Radar can detect oil on the sea by measuring changes in the microroughness of the sea surface, changes which are caused by the effect of oil on the physical surface properties of water. It is theoretically possible to calibrate the results for the thickness of the slick.

Another radar technique is sensitive to temperature, and could measure oil by determining unequal absorption and radiation of heat by surface oil and water.

Photography is a promising technique, in either the infra-red, the visible, or the ultra-violet wave lengths. Again, calibration to determine oil thickness is a possibility, but problems remain to be solved before any method is reliable.

By all estimates, by the time August had rolled around, more than one million gallons and possibly more than three million had gushed, flowed, or dribbled out of the earth and into the waters of the Santa Barbara Channel. Most of this oil was lost into the environment–sunk to the bottom of the sea, dispersed by ocean currents, or, in the case of volatile components, evaporated into the atmosphere. Some of it washed up onto the beaches. Some was collected and disposed of. Some, a very small fraction, was collected for use. As time went on, Union's ability to trap and salvage the oil improved, along with its accuracy in estimating the amount of the leak. Collecting devices were first installed on March 22. In the beginning, the variety of hoods, funnels, and tents, coupled with surface skimming devices, were able to salvage only about 10 percent of the leaking oil–about 120 gallons per day. Beginning in April, techniques improved steadily until by fall as much as 80 percent could be recovered although there were still plenty of days when things went wrong.

By the end of July, there were 14 wells in production on Union's Platforms A and B. They were withdrawing oil at a rate of 420,000 gallons per day, yielding, in the first two months of production, nearly 30 million gallons of oil. The drilling and grouting program was proceeding in parallel with the production. Six shallow grouting holes were drilled into the sea floor east of Platform A, and two more to the west. They were cased to about 150 feet and then drilled another 100 feet, to a total depth of 250 feet. A total of 24,855 sacks of cement were poured into these eight holes. Where it went, nobody knows. It gives another indication of the amazing porosity of the caprock in the area. Much of the volume may have gone to fill open fractures and fissures, but none of these could be identified.

It was clear that the seepage rate was decreasing, but no one could say with confidence whether it was from natural causes or as a result of the grouting and production programs. Then, in December 1969, the answer came unexpectedly. On

December 15, a small leak was detected in the pipeline from Platform A. This leak was quickly repaired, but the next day there was a sharp drop in the shipping line pressure, and oil contined to rise to the surface of the water. Both platforms were shut down for the repair of the shipping line leak. The repair was finished in four days, and production resumed on December 20. During the four days of the shutdown, an estimated 17,000 gallons of oil leaked to the ocean surface. More than a ten-fold increase in the leakage rate. The pressure-reduction pumping program was apparently having a marked effect. Within twelve hours after production was again underway, no more oil was to be seen rising to the surface. The second panel, the Department of the Interior, the Union Oil Company, all were exonerated. Not forgiven, but exonerated, for their controversial solution to the seepage problem.

VIII Ecological Damage: A First Look

The sea was wet as wet could be,
The sands were dry as dry.
You could not see a cloud, because
No cloud was in the sky:
No birds were flying overhead–
There were no birds to fly.
Lewis Carroll

It was the plight of the birds that stirred the most feeling. Everyone who has spent time in reflection along the California coast knows what a tremendous number and variety of birds may be seen: scurrying shore birds, wading birds, diving birds, birds that cruise above the water on the alert for a cue to dip down suddenly and skim along the surface, snatching a quick meal on the wing. Santa Barbarans love "their" birds. The oil was killing them by the thousands. Even if research should ultimately show that of all organisms in the marine environment the birds suffered the least long-range damage, it is still the birds that gain our pity. It is easy and natural to grieve for a pelican; more difficult, fot a barnacle.

The press capitalized on the slaughter of the birds. Wherever photographs of the oil spill appeared, there was a generous supply of views of oil-drenched birds, pitiful creatures already doomed but still struggling to free themselves from their black fate. One particular photograph epitomized the horror of the birds: a California murre trying vainly, desperately, to rise from a clinging pool of goo.

The Fate of the Birds

Within a day or two of the blowout, reports of dead and dying birds began to appear. On January 31, after residents saw more than 50 dead grebes along a five-mile stretch of beach near Ventura, local citizens and humane societies prepared themselves for the onslaught of victims. Union Oil established a treatment center in Carpinteria Beach State Park. Other centers were set up at humane societies and at A. Child's Estate, a children's zoo in Santa Barbara. Fish and Game Department staff, working with oil company crews, were dispatched to "problem areas" to pick up oiled birds and load them onto trucks headed for treatment centers. They were assisted by volunteers from the Audubon Society and other groups, and by countless school children, housewives, and businessmen who seemed driven to assuage their feelings of helplessness and hopelessness by helping the birds.

Oil poses a double-barreled threat to birds. There is the physical and mechanical problem of being hopelessly coated so that flight and escape are impossible. If an oiled bird remains on the water, he may drown because his feathers are stuck together and he loses his natural buoyancy, or he may die from exposure after the chill sea water reaches his skin. If he takes refuge on land, he will quickly use up his body's stored reserves and starve to death. Even a small amount of oil can prove fatal if the fouled bird preens himself and in the course of preening ingests a lethal dose of oil. The cormorant, a diving bird, was one of the first species stricken. The cormorants tried vainly to preen themselves and each other, but succeeded only in clogging their eyes and choking their beaks with scum. It has been suggested by some, but denied by others, that some sea birds are actually attracted to oil. Perhaps the darkened water resembles the upwelling of a plankton-rich current; perhaps the smoothness of the coated water is enticing. If so, the birds are lured to their death.

Treating oiled birds is a discouraging business. That much was known at the beginning. The recovery rate is vanishingly small. Still, the attempt had to be made, even if it was only a compulsive, ritualistic gesture of despair and guilt. And there was always the chance, the hope, that an effective treatment would be found.

At the treatment centers, the living birds were separated from the dead. The survivors, most of them sick, weak, cold, frightened, and in some degree of shock, were given a chemical washing. Poly-complex A-11 was most often used. It was effective in removing the oil and did minimal damage to the birds. After the bath, the birds were rinsed and carefully dried and placed in a warm room until they either recovered or (as they did with pathetic regularity) died. It took a long time–sometimes until the next molt–to restore the feathers to their normal condition and to replace the body oils lost in the cleaning process. During this period the already-weakened bird was an easy prey to respiratory disease. Added to his other problems was his frequent reluctance to eat, or the inability of his bumbling keepers to satisfy his needs. Sometimes it took only the observation that a big fish was more than a little bird could swallow to remedy the situation. At other times, the mere presence of a feeding bird stimulated a lethargic one to eat. More often, however, the traumatized birds had to be force-fed to save them from starvation.

Through trial and error, a lot was learned about the treatment and rehabilitation of oiled sea birds. Most of it came too late to benefit Santa Barbara's victims, but it sheds hope on the future. As do human patients, bird patients profit from "natural" surroundings. Cormorants and murres recover more satisfactorily if allowed to swim, so long as they are kept warm. Scoters and mergansers do best when they have weeds to hide in. The cardinal rule for grebes is, never give a grebe access to any place into which he can stick his head and then have trouble withdrawing it. If you do, he will, to the sorrow of both of you. It is important to care for the feet of a convalescing bird so that they neither dry out while the bird is denied free access to water nor become infected from stagnant dampness. If anchovies are used as feed, it is necessary to supplement the diet with thiamin (vitamin B-1) to compensate for the destruction of this vitamin by thiaminase, an enzyme present in large amounts in dead anchovies. It may be wise to provide additional thiamin routinely for all birds.

After the Torrey Canyon oil spill, overworked veterinarians and volunteers in the treatment centers resorted to a system of triage in order to be able to help any of the birds. The wisdom of this seemingly sensible move was challenged by the Santa Barbara experience. At the A. Child's Estate, each bird was permanently banded on admission and described as lightly, moderately, or heavily oiled. To the surprise of all, a bird's chances of recovery seemed to bear no relationship to the extent of his exposure to oil. Perhaps the critical factor was promptness of treatment. Perhaps some of the lightly oiled birds had deteriorated too far, by the time they were treated, to recover, while their oil-drenched brothers were found and therefore helped sooner.

During the first week of the oil spill, 234 birds were turned in at the Carpinteria and Santa Barbara treatment centers. Of these, 108 were dead or died soon after treatment. Then the pace quickened. By the end of the second week, about 850 birds had been rescued. "Rescued" from what, for what? More than 100 of them were dead when they arrived at the centers, and another 150 to 175 died within a day or two after treatment. Treatment and rehabilitation of oiled birds is a very long, difficult, expensive, and time-consuming process. If it is not done carefully and conscientiously, it would be better not to attempt it at all. Even though it hurts us to see the wretched creatures, it may be well to inquire where our efforts could be more profitably spent.

Most numerous among the victims were western grebes, followed by surf scoters, eared and horned grebes, all three species of loons, murres, and cormorants. By February 17, three weeks after the spill began, only 503 out of 1009 treated birds remained alive. As time went on, fatalities outnumbered survivals by an ever-increasing margin, so that by the end of February a mere 246 birds were alive out of more than 1600 treated.

Oil-smeared birds died on the sand at Zuma Beach, far to the south in Los Angeles County. Corpses by the dozens littered every major beach from Pismo Beach on the north and Ocean Beach at the mouth of the Santa Ynez River all the way to Point Concepcion. Beach patrols more or less quit counting the corpses. At Point Concepcion, 44 miles west of Santa Barbara Harbor, dead and dying birds lay everywhere. Ducks, gulls, loons, brandt, and miscellaneous shore birds screamed and struggled in the oil. And out on rocky promontories, the usually wary cormorants sat soggily, too exhausted or confused to fly away when approached.

Late February was nesting season for pelicans on Anacapa Island. Flocks of birds should have been picking out their nesting sites. Yet Waldo Abbott, curator of ornithology and mammalogy at the Santa Barbara Museum, reported that he had not seen a single pelican on the island. Similarly, there were only a few scattered groups of birds at Carpinteria Reef. Shores that a few weeks before had teemed with sanderlings, godwits, willitts, and curlews were now almost deserted. No ducks, geese, pelicans, or cormorants skimmed over the waves.

The oil companies optimistically reported that none of the species affected by the oil were rare or threatened. Perhaps, but that depends on your definition of "rare" and "threatened." Pelicans are considered to be endangered. In recent years, the reproductive success in some rookeries has been near zero. Many of those pelicans that survived the oil survived only to lay thin-shelled eggs laden with DDT and pesticide residues.

It is difficult to assess the bird damage caused by the oil, whether from its direct toll or from its disruption of nesting habits and sites or from its effect on the food supply. Many of the survivors took wing and flew. Those that flew away, would they return? By late summer the answer seemed to be "yes." The birds were returning to their accustomed haunts. Yet nagging doubts about their future remained.

The Migrating Whales and Their Dolphin Cousins

Along with concern for the birds came fears for the sea mammals. The southern migration of gray whales was at its height, and the usual route would lead them directly into the oil slick. There were a few reports that bands of whales were swimming through the oil. No one ventured to predict the effect oil might have on them, but everyone was prepared to hear the worst. The press did its best to give the public the sensationalism it wanted. Reporters had to rely mainly on insinuation, because there was little or no evidence of harm to the whales. Consider this report from the Santa Barbara *News-Press* of March 15:

> The California Academy of Sciences confirmed the deaths of the six California gray whales this month and said the toll compared to an average of about one dead whale . . . every other year. The latest dead whale washed ashore at Laguna Beach yesterday. Five others were carried ashore along a 70-mile stretch of northern California beach. . . . The federal government ordered autopsies on several of the creatures. . . .
>
> So far there has been no conclusive evidence linking the whale deaths to the oil disaster which began January 28.

Look at what this report actually says. It says that six dead whales were found and no one could show that they had died from causes related to the oil. The newspapers were not exactly meticulous in their reporting of details. Of the "six California gray whales" referred to above, one had died about a year before the oil spill, one had been harpooned, a third had died of pneumonia, one was a sperm whale, and one was not a whale at all, but a dolphin. The Bureau of Commercial Fisheries performed autopsies on the three animals whose cause of death was unknown, and reported that "no evidence was found of crude oil or dispersant materials in the tissue samples."

According to the Department of the Interior, between 15 and 20 gray whale carcasses washed ashore onto the beaches of California in the years from 1960 to 1968—about two per year, not one every two years, as the *News-Press* stated. At least three gray whales died in each of these years: 1961, 1963, 1964, and 1968. There was nothing remarkable in the 1969 mortality.

One dead bottlenose dolphin washed up onto the beach in mid-February. Upon examining the carcass, one vertebrate zoologist declared that the animal's blowhole was choked with oil and it had apparently smothered as a result. It is possible that the dolphin died as postulated, from a clogged blowhole, although studies of dolphin behavior make it seem unlikely. It is at least as likely that the dolphin, already dead, floated through some oil and filled its blowhole. That is, if in fact its blowhole was clogged at all; for the report was not substantiated by others who examined the animal.

About a week after the dolphin episode, the *News-Press* quoted another scientist, also of the university at Santa Barbara:

> Sea lions, porpoises, and dolphins, etc. to date, according to the Fish & Game Department, show no indication of being involved with the oil. I have gotten a pathological report on one or two dolphins [was it one—or two?] and I have determined that they are not capable of making a chemical analysis. I have this admission from their marine biology laboratory. . . . They don't have the capability of detecting the chemical residue to determine pathologically whether or not the oil caused death.
>
> They have not allowed University personnel here to make any of the autopsies. . . . The Marine Biology Laboratory at Terminal Island, California, also states that they have no capability for determining the presence of chemicals in the water. . . .

The clue may be in the second paragraph: "They have not allowed University personnel to make any of the autopsies." Since the beginning of the oil spill, university staff had felt that they had much to offer but that their capabilities were being overlooked or perhaps even intentionally shunned. They suspected a whitewash job from the government and the oil companies. They were ready to view any government report with suspicion. In the case of the dolphin(s), a negative result from a chemical assay may have one of two meanings: It may mean that there were no chemicals present, or it may mean that, if present, they were in concentrations below the threshold of sensitivity of the test. The test used would detect a level as low as 50 parts per million of a crude oil constituent. Below this level, results would be unreliable. This is not quite the meaning that the *News-Press* quotation gave to the results. It equated "no evidence of chemicals" with "we don't know how to find chemicals," as if nothing but a positive result were acceptable. It is the old problem of interpreting a negative result, but the innuendo is clear. And if traces of oil or chemicals *had* been found in the dolphin tissues—would that have proved anything about the death of the dolphin?

Whatever subtle damage the whales may have incurred, there was no mass mortality as had been feared. Dr. Ted Walker, at the University of California in La Jolla, reported that the migrating whales were arriving in the usual numbers and in the usual way at Scammons Lagoon in Baja, California. Only the young whales, born during the last season, were still in transit in mid-February. The migration of the young is slightly different from that of the adults; they steer a course close to shore, which probably kept them out of most of the oil. As for the adults, said Walker, they are alert and intelligent animals. If possible, they avoid discolored water, sometimes going as much as 12 miles off course to skirt a suspicious area. When the danger is past, they return unerringly to their course.

Seals and Sea Lions

SLICK PERILS SEALS was the large headline across the front page of April 4's Los Angeles *Herald Examiner.* Then, under it, "Oil Soaks Herd At San Miguel," and "Fur Loss Feared by Scientists." The article further explained the fears of scientists. Not their data, but their fears. Oil coated a 5 by 150 yard strip of beach to the north of Point Bennett, on the opposite side of the island from the customary gathering place of the animals. Four dead elephant seals had been found, but there was no evidence to attribute their deaths to the oil. The emotional impact of the article was strong. It would take a degree of detachment rare in those days to note that no actual catastrophe had occurred, but merely that some people feared one.

On June 13, two articles on the oil spill appeared in the sister magazines, *Time* and *Life.* The *Time* article was called "Not So Deadly." It stated that the damage done to date was not nearly so great as had been feared, although there were no predictions about long-range ecological changes. In comparison, *Life*'s article was a lurid six-page spread entitled "Iridescent Gift of Death." It claimed mass destruction of seals and sea lions on San Miguel, as a direct result of the oil spill. The article turned out to be a gross exaggeration and misrepresentation of the facts.

One morning before daybreak, a writer and a photographer from *Life,* accompanied by a zoologist from the museum in Santa Barbara and several other Santa Barbara residents, had set out on an unauthorized visit to San Miguel Island to check rumors that the seals were dying. Officially, the island can be visited only with permission from the National Park Service and in the company of a marine biologist or veterinarian who is assigned by the Navy. Filled with missionary zeal, the *Life* group had not bothered with the formality of the permit. Imagine what they would have said if some oil company public relations experts had toured the island under similar circumstances!

The visit resulted in a set of observations and photographs that appear tragic, but are impossible for anyone to interpret who is unfamiliar with the habits of seals and sea lions on San Miguel. Although the National Park Service has jurisdiction over the island, the Navy, through its Pacific Missile Range, supervises access to San Miguel. The Navy, you see, uses San Miguel as a gunnery. This may account for the bullet found in one seal corpse. Or the fact that fishermen shoot seals to keep them away from the fish and pleasure boaters shoot them for fun. But aside from the shooting, because seals and sea lions are supposedly very easily disturbed in their rookeries, human traffic on the island is kept at a minimum. The sea-lion logic that keeps timid animals living in a target area is hard to imagine.

There is no carcass cleanup program on the island, so that at any given time the collective cadavers of many months, even of years, may be seen. There was nothing unusual in the numbers of dead animals seen on the island during March and April. Sometimes it is hard to tell whether an animal is dead or living. Seals sleep soundly in some highly inelegant positions. You may have to give a seal a hefty kick to determine whether he is dead or only sleeping.

Among California sea lions, infant mortality is always high and abortions are frequent. Drowning, disease, parasites, and trampling by adults take a large toll of the young; as many as half of the pups may die before the end of the breeding season. Sea lions molt. The skin beneath where the old hair has sloughed off looks black, and the animal in general looks splotchy. The black spots could easily be mistaken for oil. The observations of the *Life* group could therefore easily be given another interpretation, one much more compatible with other observations and known facts. Most of the animals thought to be dead were only sleeping. Most of the "oil-splotched" adults were merely molting. Most of the dead infants were oil-coated mummified remains of earlier casualties, or were among the usual high proportion of aborted fetuses. Seals and sea lions were not dying at all. They were alive and well, cavorting in the oil on the sea off San Miguel.

Actually, the oil came at a fortunate time for both the elephant seals and the California sea lions. By mid-March, the elephant seal pups had already been weaned and were spared the effects of ingesting oil while nursing at their mothers' oily teats. Pupping season for sea lions is the end of May and early June. Only premature pups, which had almost no chance of survival, were born during March. Large numbers of the young of both species were examined and tagged in the spring and summer of 1969. Later sightings of tagged animals gave no indication that oil had harmed any of the animals.

Although it was widely assumed that the oil on San Miguel came from the Santa Barbara oil spill, the source actually was, and is, unknown. It is not uncommon for oil to wash up on the island, although the amount that appeared on March 19 was unusual. It may have come from a natural seep located about two miles off the island. It may have come from one of the many tankers that continually pass through the channel. Or, it may have come from Platform A. To have come from Platform A, it would have had to move against the prevailing northwesterly winds. It is possible that one of the storms that swept the area from the southwest drove the oil to San Miguel. So there is no proof that the oil came from Platform A, and there are reasons to suppose that it did not. In any case, the pollution was not severe. Union Oil and the Department of the Interior may have been guilty of certain sins of omission and commission; but they were not responsible for mass death and pollution on San Miguel.

Scientific Versus Legal Tradition

The Smithsonian Institution's Center for Short-Lived Phenomena publishes a series of "Event Information Reports" from which the following excerpts are taken. These Smithsonian reports are derived from various sources, including newspaper reports and personal communications. They describe what was known and feared about the early ecological damage in the channel, but more importantly, they illustrate two common methods of argument. We can call these methods the Legal Method and the Scientific Method. In the Legal Method, based on Anglo-Saxon tradition, anything goes short of actual lying. Inference, innuendo, appeals to the emotions—all of these are freely used to put across a point. Facts are freely selected or discarded as required. It is an effective method in courts, where it belongs, where for every view thrust on the judge and jury there is an opposite view championed with equal passion and vigor. However, the Legal Method was the one used by the press, by certain conservation groups, and by many local officials; and in their hands, it could not hope to provide complete information or a balanced outlook.

One of the common tools was the non-sequitur:

> As to fishes, etc., we have quite a number of conflicting opinions but since I talked to you last the oil has struck on 12 miles of beach above and below the city of Santa Barbara and is extremely thick and it is a disaster in the harbor.

So the beaches were a mess. What, if anything, did that have to do with the fish? Again,

> The Calfisheries Organization, an organization comprised of members of the fishing industry, will do a biological study but they already have biological information on destruction of this type, and they tell me that five years would be a limit for re-claiming. This would be the first chance for the algae and plankton balance to come back. You would also have no smelt and no anchovies which are the principal "feed fish," the bait fish, for all of the larger fish that people catch here commercially.

How convenient, to have the results before the study is begun.

Here is an example of a report that does not contain a single bit of data on oil-related damage, yet conveys the impression that an ecological disaster has occurred or is about to:

> After a survey least week of the oil polluted waters off Santa Barbara, Ian McMillan appraised the damage as being far greater than was indicated in official reports. Populations of rare marine mammals on the four Channel Islands are of particular concern. Some 4000 sea elephants which range this area have their young at this time of year. The only northern fur seals south of Alaska are represented in a breeding colony on one of the islands. This population of some sixty adults amazingly appeared here only in the last two years and last season was reported to have produced ten young. Sea otters have been seen in the area within the last year. It is the aim of our group to see that these precariously situated mammal populations are given full attention for whatever impact they may suffer from the encroaching oil. The main ecological effects of the pollution will come through damage to the basic food supply–that which makes up the start of the food chain. Its effects on these foundation elements will undoubtedly be disastrous and will take years to assess.

Telling only half the story can be effective:

> G. Raymond Arnett, director of the State Fish & Game Dept., said last night that the contamination has had no observable influence on fish life in the channel, but that "long-term effects on the marine environment" are feared. The Fishermen's Assn. reported, though, that catches in the Channel are down 90 percent.

There is no conflict here. Arnett said there were fish in the channel; the Fishermen's Association said they were not being caught. The Fishermen's Association neglected to say that throughout the month of February commercial fishermen were for the most part unable to take their boats out of the harbor. The harbor was completely closed during a large part of the time; at other times, the harbor master allowed ships to enter and exit only during brief periods which he himself designated. Many of the fishing boats had been damaged by oil and were not ready for use. From the immediate view of the fishermen, whether or not there were fish in the channel was academic because the oil would ruin their nets and other gear if fishing were attempted. We can imagine the version of the Fishermen's Association report that might be released by the "Fishes' Association," whose scaly members frolicked unmolested in the channel: "Commercial fishermen down 90 percent in the month of February!"

Assumptions, scrupulously acknowledged as such, were another strategy of the Legal Method:

> There were numbers of birds sitting along the beach. We could not determine whether they had been involved with the oil or not. *We could only assume they had.* (Italics ours.)

Doomsday predictions had their place:

> Dr. Carl Hubbs, a marine biologist at the Scripps Institution of Oceanography, La Jolla, predicted that the oil would kill all marine life along 20 miles of coast.

The overall impression for the unwary was that the end of the world had come, prematurely but indisputably, to Santa Barbara. Of course, this is what was intended. The newspapers were full of this kind of reporting because it sells. Pressure groups of one kind and another used these tactics in the hope that emotion would overrule reason long enough for them to win their drive to get oil out of the channel. Facts, after all, are exceedingly dull and much more quickly forgotten than is the silly, innocent face of a baby sea lion.

The Scientific Method, in contrast to the Legal Method, places its hopes on the presentation of data and careful observation. Fact after fact is piled up until the weight of the accumulated facts renders the conclusion indisputable. Some in Santa Barbara sought out the facts:

> We were out in Santa Barbara Channel, and we could find no noticeable effect on the fish eggs and larvae in the mid-water under the oil. The expected groups were there—the ganid fish, the cupiads, the hake, anchovy and rock fish. Their larvae was all there in expected amounts. In the Santa Barbara harbor, of course, you have serious damage in the inter-tidal zone. Oil is continuing to come up in this zone and causing serious damage especially to the organisms there (abalone, mussels and vegetation). . . .

Not all of the facts presented by practitioners of the Scientific Method were reassuring:

> The plankton probably were affected in the enclosures—the harbor. We couldn't see anything in the plankton out in mid-water. We had light attenuation under crude oil up to 99%—the surface layer of oil cut down the light so that only 1% remained but that was in a small area. However, the fact that the oil is moving in one direction and the water in another means that such low light levels would exist for only a short period of time.

A Legal Method Man might have stated that 99 percent of the usual light was cut off by the oil, and then asserted that since light is required by phytoplankton for photosynthesis and since all life depends on photosynthesis, the ocean was about to die and we might as well all give up.

There was good news and bad news, tempered by uncertainty and caution:

> We have a pretty good idea from a previous 4-year study as to what the environment was before. . . . We took some plankton tows right around the vicinity of the leaking platform. It is safe to say that the immediate area is a mess; the slick is on the surface. We were able to make collections of live organisms right under the oil slick. . . . If the plankton organisms in the

> vicinity of the oil leak itself are in any way contaminated, they might serve as the transmitted food chain contaminant to larger organisms. . . . We found things like larval fishes and the young of the year which means that in the case of a few fish, they have successfully spawned. Bottom fishes seem to be normally abundant.

Summary of the Immediate Damage

Overall, the immediate ecological damage was negligible–nothing like what previous experience with oil spills led people to fear. Although large numbers of birds were killed, by summertime many populations already showed signs of recovering. Most species can survive a one-time mass slaughter. It is destruction of their habitats and elimination of their food supply that pose the ultimate threat to survival. Even the direct bird kill itself was less than had been feared. Perhaps it was because there was no widespread use of detergents, as there had been in earlier oil spills.

Whales and dolphins avoided the oil. Seals and sea lions played in it happily. Oil was no novelty to them. For as long as there had been seals and sea lions in the channel, there had been oil, and the animals had romped in it with apparent immunity. These large sea mammals are unlikely to blunder into the oil and be caught unawares.

Most of the adult fishes, too, were able to avoid the oil. A large fish kill could not have passed undetected. As none was reported, it is almost certain that none occurred. There is excellent communication among members of a school of fish, which helps them to stay together and to shun unfriendly waters. Some fish swam away from the troubled waters, to return on a better day. Others, which normally live near the surface, went deeper when the oil came. Still others did not change their habits at all, but were not harmed by the oil.

Some damage could definitely be laid to the oil. The dead mussels, rotting in blackened tide pools. The yellowed, decomposing surf grasses, bathed alternately in fresh oil and oil redistributed when the high intertidal rocks were cleaned. Red algae–now no longer red, but white and brittle. Several species of brown and green algae, temporarily affected but soon to recover. Barnacles–many smothered, some still surviving but coated with oil long after natural processes had cleaned the surrounding rocks.

At first there were fears for the kelp beds, which trapped large amounts of oil. This capacity of the kelp turned out to be a nuisance, but the plants themselves suffered no harm. The kelp was coated with a mucoid slime, which prevented oil from coming into contact with living plant tissue and also prevented oil from clinging to the plants. (A similar mucoid layer protected many of the smaller algae.) Intermittently, however, depending on the vagaries of the wind and waves, large quantities of loosely held oil were swept free from their kelp prison to contaminate or recontaminate beaches and coastal waters. For this reason, many advisers recommended cutting the kelp beds that were heavily fouled with oil.

Between early February and mid-August, several research vessels made repeated visits to stations established soon after the oil spill began. Each survey brought encouraging reports. All species of fish appeared to be spawning and developing normally, and their overall abundance and distribution corresponded to the long-term average for the date and the area. Healthy specimens were taken repeatedly from beneath the oil slick. Certain algae at Santa Cruz Island, heavily coated with oil in February, had disappeared by April; but in August, young plants were again growing vigorously. Only the surf grasses, hardest hit of all the plants, were apparently in for a long-term recovery and recolonization period.

With the early results in but the final tally yet to come, those who had thought the end of the world had come began to express hope. Scientists, with guarded optimism, reaffirmed that they had found negligible damage to the marine ecosystem, but warned of the dangers of chronic pollution from the continuing leak. There were those who thought they detected subtle changes that boded ill—like the surf grass, which was not recovering as it should. But overall, what seemed at first to be sure disaster had miraculously turned out to be little more than a terrible mess.

The reaction of the press and many conservationists to the situation in Santa Barbara probably did the conservation movement as much harm as good. The kind of sensationalism we have examined in this chapter is readily spotted by the policy makers in Washington, who are learning to discount much of what environmental alarmists say. The whales, after all, did *not* die.

After the birds were cleaned up and the oil spill gave way to other front page tragedies, people went on to worry about other things. Very few stayed to hear the end of the story. But the worst part of the problem is not, as Fred Hartley phrased it, "the death of a few birds." The environmental studies described in the next chapter provide the only rational approach to assessing the true nature and extent of the problem. Armed with data like these on the condition of our coastal waters, the environmentalist should be able to shake the complacency of Washington and the nation so that we can direct our energies to where they will be most effective.

IX *The Toll*

Always spend a penny
as if you were spending a
dollar
and always spend a dollar
as if you were spending
a wounded eagle and always
spend a wounded eagle as if
you were spending the very
sky itself.
Richard Brautigan

We have already said a great deal about marine birds, fish, and mammals, but very little about less familiar animals of the shore and sea. Only a casual examination of rocky shores and tide pools reveals an amazing display of fascinating animal and plant forms. From Point Concepcion to San Diego, all of these, to some degree, were menaced by the oil spill. A mere recitation of the casualties and survivors would have little meaning, in the absence of some acquaintance with the creatures themselves. The two sections following will introduce some of the characteristic organisms from the main taxonomic groups that are found on the Pacific coast.

The Food Chain

The story of life in the intertidal region begins with the tiny plants called diatoms, which are the very base, the beginning, of the marine food chain. These beautiful and varied microscopic cells are able to live and grow by building their own food from materials readily available: sea water, with its dissolved minerals and carbon dioxide, and energy from the light of the sun. The process by which they do this is called photosynthesis, a kind of putting-together-by-means-of-light.

Diatoms and dinoflagellates (another group of minute, photosynthetic organisms) make up a large part of marine plankton, which is all the tiny plant and animal organisms that swim weakly or drift about in the water. Diatoms and other small algae make up the *phyto*plankton. A number of major animal groups are represented in the *zoo*plankton. Chief among the protozoa are the foraminifers, which contribute both to the plankton and to the benthic organisms (bottom-dwellers). Most of the foraminifers are barely visible to the naked eye; a few may be as large as a pinhead. Countless numbers of these tiny individuals secrete their many-chambered chalky shells which they leave behind when they die. It was deposits of foraminifer shells that created the famed chalk cliffs of Dover, England, and the chalk beds of Mississippi and Georgia.

Also numbered among the plankton organisms are the eggs, larvae, and juveniles of many larger animal species, as well as some very tiny crustaceans. These crustaceans, largely copepods ("oar-feet"), feed on diatoms and dinoflagellates with perhaps an egg or two that they filter from the water. Beyond the copepods, the food chain becomes diffuse, with everything being likely to be eaten by anything. Much remains to be learned about who eats whom in the ocean.

The success of each higher order of animal depends on the condition of the ranks below, so that, far from being the master man supposes he is, man, at the top of the food chain, is one of the most vulnerable species of all. Along with other predatory mammals and birds, we are sustained by the ranks of smaller creatures behind us. We have our parasites, as well; and the dead body of a man or a whale furnishes food for countless smaller creatures which eventually return many of the building blocks to the sea, to the diatoms, where the process begins again. This interdependence of hierarchical levels is the key to our often repeated statement that one of the gravest dangers to our species or to any other from pollution is the disruption of the food chain.

Plants and Animals Between the Tides

Life abounds in the intertidal region, the coastal strip between the highest high tide and the lowest low tide. Compared to this region, the open sea is a barren desert. The coastal region, too, is the most vulnerable to damage from oil pollution and the most frequently assaulted. Here, the sands and rocks are alternately covered and exposed by the advance and retreat of the tides–each receding tide potentially capable of leaving a coating of oil for the next high tide to wash away. While the coast gets its twice-daily oil bath, oil on the open sea can be rapidly diluted and carried away. The coastal region is most threatened because it is here that the effluent from our land-based industries is dumped, here that vessel traffic is heaviest, and here that much production, use, and transportation of petroleum occur.

High up on the rocky beaches, where the sand is wetted only by the spray from the pounding surf, small dingy snails called periwinkles glide along, scraping microscopic plants from seemingly barren rocks. In the blind, unconscious way in which such changes occur, some species of periwinkles seem to be trying to give up their marine life in favor of the land. They show a pronounced distaste for the sea, which they still grudgingly need, staying close enough only to wet their gills once in a while. Periwinkles and their other snail cousins belong to the large group of animals called molluscs, a group which also includes limpets, chitons, clams, oysters, and octopi.

In the company of the periwinkle may be found some small crustaceans, the pill bugs and sand fleas. Some of the pill bugs, like the periwinkles, may be pursuing an evolutionary course away from the sea, as they scutter over the rocks far from the water. Sometimes they follow the receding tide down to the intertidal area in search of food, but they abhor wet feet and stay out of reach of surf spray. Other pill bugs live lower in the intertidal zone, but they all show a distaste for prolonged bathing.

A few flatworms may be found lurking under rocks in the lower part of the uppermost region, where they stay damp at all times and are occasionally thoroughly wetted by the high tides. Barnacles are almost ubiquitous in the intertidal region, only the species and numbers of individuals varying with the habitat. Barnacles are known to everyone by name, but their fascinating life history is unknown to most. They are crustaceans. As crustaceans, they are broadly related to copepods, pill bugs, sand fleas, shrimps, and lobsters. They are hermaphrodites and are usually self-fertilized. Any other reproductive scheme would be incompatible with their immobile, sometimes almost solitary adult lives. Eggs and larvae are retained within the parents' shells, to be discharged into the water as tiny, free-swimming "nauplii." The nauplius is a preposterous one-eyed, one-shelled baby with three pairs of legs. Through several transformations he becomes an equally preposterous "cypris," with two eyes, two shells, and six pairs of legs. Now he seems inclined to settle down. His antennae become suckers, by which he fastens himself head-first to his selected site. He secretes an amazingly strong and durable cement, which is being studied by scientists today in the hope that it will lead to a more permanent and satisfactory cement for human dentistry. Now the cypris-barnacle begins work on his limey adult shell. He is blind and sessile. When covered by water—perhaps for only a few hours each week, in the highest zones—he will open his doors and extend his feathery legs, rhythmically sweeping the water to breathe and to catch the zooplankton on which he feeds.

The limpet, related to common marine snails, is another interesting animal. You can find him in action at night, when he extends his foot, raises his shell from its daytime roosting place, and glides off in search of the food which he scrapes from the rocks with his filelike "radula." To catch him you must take him by surprise, for at the slightest provocation he jerks into his shell and sucks himself so firmly against his rock that you will break his shell before you can pry him loose. Someone once measured that it took a 70-pound pull to remove a limpet having a basal area of one square inch.

The next lower coastal region extends approximately from the average high tide level to the average of the high low tides. There is considerable variation in the tides, depending on local conditions and the phase of the moon. (This cycle plays a role in the natural cleaning of contaminated beaches.) High tides are higher and low tides much lower at times of the new and full moons, when the pulls of the sun and the moon reinforce each other.

In this region of fairly permanent tide pools, life is much more varied and abundant than in the higher region reached only by the spray of the surf and the very highest of tides. Here, barnacles are larger and more numerous. Other species of periwinkles, less daring in their terrestrial exploits, replace the ones found farther from the sea. Flatworms flow along damp rock surfaces, propelled by many hairlike, leglike cilia. You cannot pull one from his rock without damaging him, but he is a ticklish fellow and you can tickle him off with a feather.

The renowned hermit crabs hustle around the tide pools in their secondhand shells, fighting among themselves in a ritualistic way that assures that no one ever gets hurt. The hermits, having lost the ability to build their own homes, must periodically, as they grow larger, go house-hunting among the empty shells that litter the coast. If a new shell, perhaps the shell of a turban snail, passes a critical inspection, the new roomer will suddenly make up his mind and zip inside. That is home, until he outgrows it.

Many of the bluegreen and some of the green algae begin in this region between the tides, and extend on out to sea. The alga *Fucus* and other large algae serve as bases for attachment for sessile animals, provide shelter for other animals, and protect many creatures from desiccation and from the direct rays of the sun.

Other animals live on and under tide pool rocks: the tiny porcelain crab, for example, that scurries frantically for cover when his stone is overturned, perhaps shedding a leg or two in the process. This ability to cast off an appendage "at will" is shared by many invertebrates. Coupled with an amazing capacity for regeneration, this device provides escape and life for many an animal caught by a predator or pinched between two rocks.

Brittle stars live under the tide pool rocks, with their close relatives the starfish and a few anemones. Some starfish have carried the business of regeneration so far that not only can a starfish replace a ray lost in battle, but the amputated stump can grow itself a whole new starfish!

Continuing down as far as the average low low tide, we find a region that is uncovered less regularly and for shorter periods than the adjacent region. Here plants flourish, providing food, protection, and anchoring places for countless animals. Surf grasses, true flowering plants, are common. Anemones, animals related to corals and probably so-named because of their resemblance to flowers, grow on exposed rocks in such profusion as to render the intertidal rocks squishy.

When closed up and covered with tiny bits of rock and shell, they can scarcely be distinguished from the underlying rocks. Anemones are exceedingly long-lived. Some, reportedly, have survived in aquaria for more than 70 years.

On protected rocks in the low tide region are the large red, yellow, or purple starfish called "sea bats" because of their webbed rays. There are also small starfish and gaily colored sponges. Nudibranchs (sea slugs, shell-less snails) walk upside down in the tide pool, hanging from the water side of the air-water interface. They spin out a path of slime as they go, to lessen their chances of breaking the surface film and tumbling to the bottom. The census could go on and on. Ribbon worms, scale worms, tube worms, ghost shrimp. . . .

Last comes the low tide region, uncovered only by extremely low tides. Here the creatures are not so finely tuned to the rhythm of the tides. Here, too, more species may be found than at any other tidal level. Only here grows the green anemone, a beautiful large, solitary animal which owes its startling color to the green algae living symbiotically in its tissues. Starfish are numerous: many multirayed, some with a dozen arms or more. Sea cucumbers, starfish relations, may be found indulging in their disconcerting practice of eviscerating themselves when disturbed. It looks fatal, but they quickly regenerate a new set of internal organs. Their spiny sea urchin cousins live here too, paralyzing prey and predator alike with a poison injected from a specialized appendage, and nibbling on the alga *Ulva,* "sea lettuce." Sea lettuce is also the diet of choice for the abalone, prized for his meat, his shell, and his "pearls." The chiton, lesser-known cousin of the snail, moves about on his long, flat foot.

Scallops inhabit the deeper waters of the Santa Barbara Channel and a few of them venture into the tide pool area. One species frequents the kelp beds and, until the oil spill, lived quite successfully in the vicinity of Platform A. The scallop shares one peculiar trait with his relative the octopus. That is a slow, inefficient method of forward locomotion used when the animal is searching for food, combined with an exceedingly rapid and efficient backwards escape mechanism. He zooms backwards by clapping the two valves of his shell together, producing a jet effect with two streams of water that squirt through holes in his mantle.

A species of small octopus is found beginning at Santa Barbara, its range extending southward. We may note here that the fauna of northern and southern California are distinct from one another, and for many species, Santa Barbara represents either the northernmost or the southernmost limit. These species are rather precariously situated in Santa Barbara, under stress and competition from species with overlapping ranges. For some species, including several molluscs, reproduction in these circumstances may be "normally abnormal," so that some findings after the oil spill were difficult to interpret. There was a lack of information on the normal occurrence and reproductive habits of animals in and around the channel.

One last animal is of importance, the echiuroid worm *Listriolobus pelodes.* Dense beds of these burrowing worms, which are similar in many ways to the segmented worms, are unique to the Santa Barbara coast. They inhabit mud flats that are exposed only at very low tide. Most echiuroids are accomplished diggers, constructing U-shaped burrows with two openings to the top. *L. pelodes,* however, is indifferent toward digging; he merely scoops out a shallow burrow for himself in which he sits, spoon-shaped "head" held erect in the water, flat side oriented against the current. He appears to filter food from the water as it is swept by.

The Nature of Petroleum

To understand what happened to these animals, we must know something about what oil is. Crude oil is a complex mixture of organic compounds, one of the most complex natural mixtures to be found on earth. The components of the mixture are predominantly hydrocarbons, or compounds containing only carbon and hydrogen. But because of the unique chemical properties of carbon, an almost infinite variety of molecules may be built from just those two elements, carbon and hydrogen. Many refined petroleum products are 100 percent hydrocarbons. In some of the other molecules found in crude oil, perhaps only 2 or 3 percent, atoms of sulfur, nitrogen, or oxygen may also occur.

Three basic types of hydrocarbons occur naturally in crude petroleum while a fourth may be created by "cracking" during the refining process. The first type of compound (the paraffins or alkanes familiar to students of organic chemistry) has a chainlike structure, either straight or branched, in which each carbon atom is linked to four other carbon or hydrogen atoms. All the available bonds of the carbons are thus used, and the compound is "saturated." The names of the simplest members of the straight chain series are familiar: methane, ethane, propane, butane, pentane . . . octane. . . . Straight- and branched-chain hydrocarbons of this type constitute a large fraction of crude petroleum and predominate in gasoline and some fuels. Beginning with octane, the eighth member of the series, their solubility in water is negligible. This does not mean, however, that they do not mix with sea water. Through wind and wave action they may be dispersed as tiny droplets in the sea, or they may even hitch a ride to the bottom on clay or other particles and thus contaminate the bottom sediments.

In the second type of compound, which is also saturated, some of the carbon atoms are joined into rings of from five to seven members. These are the alicyclic compounds—"cyclic," because of the rings. The simplest compounds of the group are cyclopentane, cyclohexane, and cycloheptane, with five, six, and seven carbons, respectively, in the rings. Other chemical groups may be substituted for the hydrogens in these molecules.

The third type of compound, like the alicyclics, contains one or more rings. The carbon atoms in these rings, however, share their bonds with only three other atoms, rather than four. The simplest compound of this type is the familiar benzene. Other common organic solvents chemically related to benzene are toluene, ethylbenzene, and xylene. Many of the larger molecules have been implicated as potent carcinogens (cancer-causing agents) in experimental animals. Many are known to be acutely toxic to man or other animals.

The fourth type of compound is a class of unsaturated, straight- or branched-chain compounds called olefins or alkenes. They are produced during the refining process, when through "cracking," large molecules are broken down into smaller ones that make up the more desirable light petroleum fractions.

In the process of refining, the components of natural crude oil are separated into fractions according to their boiling points. We get numerous products, including gasoline, kerosene, several grades of fuel oil, and a residue that yields asphalt. This separation is accomplished in a fractionating tower, which is nothing more than a huge distillation apparatus in which the crude oil is heated and the fractions are condensed and drawn off at different levels in the tower, low-boiling at the top, high-boiling at the bottom. Some of the constituents of crude petroleum are gases at room temperature. Others have boiling points well above 500 degrees Centigrade. Because different types of compounds are concentrated in the various petroleum fractions, much has been written about the relative toxicity of different petroleum products. Volume for volume, crude oil appears to be less toxic than, for example, fuel oil. However, with the exception of the few compounds created by the refining process, all of the chemical constituents in the refined parts are also found in the unrefined whole. If the part can be deadly, so can the whole.

The Fate of Oil in the Sea

When crude oil or a petroleum product is discharged into the sea, it begins at once to undergo a series of changes. Many of the low molecular weight, volatile components are hit-run killers, doing their damage quickly and then vanishing from the scene. These small molecules are acutely toxic to many organisms, but the rapid rate at which they evaporate removes a large percentage of them from the marine environment in a short time. (They still exist in the atmosphere, where they join the astonishing quantities of hydrocarbons escaping into the air every time you fill the gas tank of your car. We have barely begun to concern ourselves with the ultimate fate of these pollutants. Do they return to us in rain, enter the rivers, and find their way back to the sea?) Since evaporation is a surface phenomenon, residual volatile components of a heavy oil may remain for an extended time on the sea, trapped within thick streams or globs of oil. Some of the small molecules may be carried to the sea floor on silt particles. As the sediments move, contamination spreads along the ocean bottom.

The heavier fractions of crude oil are more persistent. They contain the slower-acting poisons. But even these heavy components are not immortal. Oil from an oil spill does not remain in the water or on the beaches forever. What happens to it? Much of it is degraded by marine bacteria. Although we are still largely ignorant about the microbiology of the sea, we have identified more than one hundred species representing thirty genera of marine bacteria which metabolize petroleum. These bacteria are most numerous near the shore, where their populations are rapidly increasing as coastal pollution from sewage and industrial sources increases. In the bottom mud, or where sewage enters the sea, there may be more than half a billion bacteria per liter of sea water. Far off shore, the number drops to a few thousand per liter, or fewer. Here, the numbers are limited by the available food supply. When oil suddenly appears in an environment where food previously limited the population, the petroleum-metabolizing bacteria begin to feed and multiply rapidly. After the food is exhausted the population dies back to its original level, the level which can be supported by the basic 5 or 6 parts per million of organic matter in the sea.

The presence of petroleum in the sea increases the "biological oxygen demand" that is of concern to microbiologists, engineers, and ecologists. This is a measure of the oxygen needed to oxidize all the organic material in the water. A high biological oxygen demand is a consequence of water pollution by sewage and other organic products, like petroleum. It may result in depletion of available oxygen supply, to the detriment of desirable organisms.

Bacterial degradation of petroleum may be a long, slow process, especially if the oil is left untreated. Part of the rationale behind chemical treatment of spills is to disperse or emulsify the oil so that a relatively large surface area is open to bacterial action, or to sink it to the bottom where the bacterial population is greatest. Microorganisms can only nibble at the outer layer of a huge glob of oil.

Is the Cure Worse than the Disease?

In earlier chapters, we touched upon some of the methods for cleaning up an oil spill, with their advantages and drawbacks. Oil companies are working very hard on this technology, and new methods appear almost daily.

Ten thousand truckloads of straw were tossed onto the sea and beaches at Santa Barbara, to trap the oil. The oil-soaked straw was then raked up and burned. Don't laugh at this Stone Age antidote for an Industrial Age poison. It was just about the most effective method anyone came up with. Other sorbants have been used, both natural and synthetic, including sawdust, rope, bark, "Ekoperl," polyurethane foam, polypropylene fiber, waste paper, glass wool, vermiculite, coffee grounds, and rayon floss. Oil has been congealed with plastic foam, plastic film, and nylon agglutinants. It has been gelled with molten wax and soaps. It has been sunk with sand, brickdust, cement, volcanic ash, talc, crushed stone, and slaked lime. But straw remains one of the best materials for collecting spilled oil and removing it from water and beaches. The main problem is disposal of the straw.

Salvage of spilled oil through removal by pumping or skimming would be the ideal solution, but technology has not provided a device that can deal with a large spill. Similarly, booming the source seems an obvious solution. Theoreticians have calculated page after page of equations which describe the properties of an effective boom in terms of the conditions of wind, waves, and current it would have to withstand. But we have not built a boom that is effective in open water.

Detergents, we learned from the Torrey Canyon experience, are much more deadly than oil itself. There are both an immediate toxic effect from the solvent and a residual long-lasting poison in the detergent. In addition, there is danger of enhancing the damage done by the oil. Oil is normally repelled by living tissues, but the oil-detergent complex can readily come into close contact with the surfaces of plants and animals. This, after all, is the very function of the detergent: to alter the surface properties of oil and water, so that they may mix.

In Santa Barbara, the traditional detergents were not used. But other "detergent-like" materials, which had the same effect on surface properties, were used. In laboratory tests, these materials varied from slightly to moderately toxic to marine organisms; but there was no evidence that, as actually used on the oil spill, they harmed marine life. Although the theoretical dangers existed, we did not observe any damage.

Steam or hot water cleaning of rocks did more harm than good. Besides killing the plants and animals that lived on the rocks, it was not even effective as a cleaning method. Oil washed down into the lower intertidal regions to contaminate and recontaminate the tide pools, and then was washed right back onto the rocks at the next high tide.

Perhaps the most that can be said about the materials and methods used for cleaning up oil spills is that it is easy to describe the properties of the ideal agent or treatment for each type of slick. But finding or inventing this perfect product is quite another matter.

Three Oil Spills, Three Outcomes

It is difficult to evaluate the effects of an oil spill. All that we know with certainty is that the variety and quality of marine life is decreasing in proportion to the increase of pollution, and that one of the major pollutants of the sea is oil. The difficulty lies partly in the fact that each oil spill is a unique event, involving a particular type of oil, a particular environment, and a particular set of remedial measures. Recovery from an oil spill may *seem* to be amazingly rapid and complete, while marginal effects, subtle changes that can be detected only on close inspection or on laboratory investigation, are overlooked. Or we may be so impressed by the magnitude and drama of the major disasters that we forget about the steady deterioration of all our coastal waters.

The situation is further complicated by the paucity of comparable studies of the resulting damage. Sometimes there is very little background information on the "normal" ecology of the area. Sometimes it is almost impossible to separate the effects of the oil pollution from accompanying effects. In Santa Barbara, we have seen that record winter floods had already made a shambles of the area before the pollution occurred. Fresh water can be more deadly to marine life than oil. Tons of fresh water had poured into the salty sea, bringing with it pesticides and silt and debris that disrupted the marine habitat from the surface to the bottom mud.

Nevertheless, it is instructive to compare, so far as possible, the effects of three well-documented oil spills: the Torrey Canyon disaster, a small oil spill in Buzzards Bay, Massachusetts, and the Santa Barbara oil spill.

The wreck of the Torrey Canyon at Easter-time in 1967 caught everyone off-guard and unprepared. Upon reflection, it seems that this tragedy, like most tragedies in history, could so easily have been prevented at so many steps along the way. But perhaps that is the essential nature of tragedy.

After the oil was spilled, thirty million gallons of it, the prime concern became preservation or restoration of the amenity value of coastal resorts. How mixed up our values must be, that we will gamble our stake in the future for the transient pleasure of a holiday swim! *Ten thousand tons* of detergent fluids were dumped frantically on fifteen thousand tons of oil on the Cornish beaches. As the tides rolled in and out over the treated beaches, they swept an expanding front of poisoned water along the coast and out to sea, over the surface and down to the sea bed. Animals living below the low tide mark were most sensitive to the detergents, but their dosage was mitigated by dilution of the detergent in open water. The intertidal animals, although relatively resistant, were often exposed to high concentrations of detergents for several hours at a time.

Laboratory tests and surveys of rocky shores revealed the expected variation in tolerance among species, a phenomenon which makes it impossible to define a "safe" level of detergent in the sea. Many algae, especially those that thrive in brackish water, were quite resistant. Plants in general suffered less mortality than did animals. Of all the animals, only the sea anemones and the high intertidal periwinkles escaped massive destruction: the former, because of their tolerance to both oil and detergents and the latter, because they lived for the most part beyond the reach of the pollution. In many small localized areas that were heavily oiled and heavily treated, all animal life was wiped out.

Britain's handling of the pollution crisis aggravated the already somewhat tense relationship between the British and French governments–a relationship which, for "friendly" nations, had suffered several cool spells in recent years. By indiscriminate dumping of detergents on 100,000 tons of oil, Britain created a repulsive oil-water emulsion (scathingly referred to as "chocolate mousse") of perhaps as much as 600,000 tons total weight. Much of this mousse mess washed across the English Channel to the French coast, where it presented more of a problem than the oil alone. Besides being enormously increased in bulk, it was much more difficult to remove from the rocks and beaches, it sank deep into the sand, and it was readily spread by the wind and waves. The French wisely emphasized physical removal of the oil, even going to the extreme of setting armies of scrubbers to work on the rocks with brushes.

The immediate biological effect of the Torrey Canyon oil spill was decimation of many invertebrate populations and some species of fish as well. It is known that a minimum of 10,000 birds was killed. The true number, which includes the hordes killed at sea whose bodies were not recovered, can only be guessed at. This animal mortality was followed by an overgrowth of marine plants of many kinds. The plants were able to proliferate because they had suffered less from the pollution than had the animals, and they were then able to multiply unrestrained by the limpets and other grazers that normally kept the populations under control. Even so, recolonization by animals from unpolluted oases in the marine wasteland began almost at once, slowly at first but then more rapidly. By the time of the Santa Barbara oil spill, recovery of most of the polluted areas seemed essentially complete.

In contrast to the massive Torrey Canyon spill of 30 million gallons, from which recovery seemed miraculous in spite of the mismanagement of the cleanup, a relatively small spill in Buzzards Bay caused widespread, long-term damage. On September 16, 1969, the oil barge Florida dumped some 100,000 gallons of Number-2 fuel oil almost at the doorstep of the Woods Hole Oceanographic Institution. At once, Woods Hole scientists mapped out plans for immediate and long-range experiments and observations.

A drastic mortality of marine life took place within a few days of the spill. Up to 95 percent of the animals recovered from bottom samples were either dead or dying. Similar destruction occurred in tidal rivers and marshes. Then, within days, most of the dead animals decayed and things began to look normal again, to the casual observer. There is a hidden danger here, a danger in equating "everything looks normal" with "complete recovery has taken place." A careful look by Woods Hole scientists showed that everything was far from normal.

Certain animals are sensitive indicators of pollution: some, because they are especially vulnerable, like some of the tiny shrimp, and others, because they are highly resistant. The segmented worm *Capitella capitata* (poetically called the "sludge worm") is notorious world-wide as a characteristic inhabitant of heavily polluted waters, where it thrives in the absence of its more delicate predators. This worm was normally rare in Buzzards Bay, but it became abundant after the oil spill. Although all visible evidence of oil was gone within a short time, contamination spread along the bottom of the sea for more than eight months after the spill. Movement of bottom sediments was probably enhanced as a result of the killing of bottom-dwelling organisms which, in life, tended to hold the sediments in place. As the contaminated sediments migrated away from the site of the original pollution, a wave of death followed close behind the leading edge. Apparently, we cannot count on evaporation and bacteria to remove or degrade the petroleum we spill on the water. This light grade of fuel oil "should" have evaporated quickly and "should" have been readily degraded by microorganisms. Yet it was thoroughly mixed through the water and carried to the bottom, where it was trapped; and for some reason, after eight months, bacterial degradation had barely begun in many of the most heavily polluted areas. One year after the spill, bacteria had made definite inroads on the oil, but many of the most toxic components still remained. One and a half years after the spill, free-swimming larvae were still being killed when they settled on the polluted sea bottom or in polluted marshes.

The oysters, clams, and scallops that were not killed outright were tainted, and contaminated grounds were closed in 1969. In 1970, contamination was worse than in 1969 and additional beds, beds that had been harvested in 1969, had to be closed. Unlike many pollutants (from sewage, for example), hydrocarbons are not washed out when the shellfish are placed in clean water. It appears that, once contaminated, the animals may be contaminated for their lifetime. Blue mussels that were juveniles at the time of the spill were sexually sterile the next season.

Eighteen months after the oil spill, only the most pollution-resistant organisms had become re-established in the most heavily polluted areas. Oil was still easily detected chemically in the bottom sediments. Things had most definitely not returned to normal. Yet this was only a small spill, a fraction of a percent the size of the Torrey Canyon spill. Why was it so vicious? Why did we escape so lightly from the Torrey Canyon fiasco, even after we added the insult of detergents to the mischief of the oil? Perhaps the results of the two spills were not really so different, after all. Perhaps we got different answers because we asked, basically, different questions. We cannot turn back the clock and get another chance to do carefully controlled experiments. But we can look at what happened in Santa Barbara and try to make some sense out of the confusing picture.

The most thorough study of the ecological effects of the Santa Barbara oil spill was published by the Alan Hancock Foundation of the University of Southern California. Unfortunately, this study was greeted with feelings ranging from skepticism to disapproval. It was funded by the Western Oil and Gas Association. To the layman who did not understand or believe in the ethics of research or the vigilance with which the scientific community polices itself, it seemed likely that the university had sold out to the oil interests. (This is not to say that there are no moral issues in science. But there is a great difference between questioning the value or the morality of doing a piece of research at all and questioning the integrity of scientists in reporting their results. Occasionally, a scientist is dishonest. It usually does not take his peers long to expose him.) When the report came out, suspicions seemed to be confirmed. Everyone knew intuitively that the oil spill had been an ecological disaster, yet the scientists were saying otherwise. The results must therefore be wrong. But a careful reading of the report reveals little that can be criticized, unless it is the basic lack of fundamental information on the ecology of the channel. This is a deficiency for which the scientists participating in the study must be allowed to share the blame with everyone. We all failed to encourage, support, or do the background research before it was too late.

As the crude oil from Platform A washed onto Santa Barbara's once-proud beaches, no one could predict with certainty what its fate would be. The oil spill occurred after the time when most of the beaches had been cut back in the annual cycle of erosion and deposition of beach sand, so there was little chance for natural removal of oiled sand to take place. Any oil not removed by beach crews was buried later in the season as new sand was deposited. Pessimists warned that each year for uncountable years to come, as the anniversary of the oil spill drew near, the annual cutback would re-expose layers of oiled sand. In fact, however, by February 1970, most of the buried oil had migrated to the surface and had been washed away. It is thought that this migration of oil accompanied movements of the beach water table. Some beaches, those southeast of the Ventura and Santa Clara Rivers, had been abnormally built up from the floods. The cycle was out of phase south of these rivers. Cutting back occurred late in the season, after the oil spill, and these beaches were soon cleaned by natural processes. Oil deposited on the beaches did not at any time alter the normal movement of beach sand.

The winter storms swept more than 70 million tons of flood sediments into the eastern end of the Santa Barbara Channel. This means that for every gallon of oil released during the oil spill there were *14,000* gallons of silt washed into the channel. Imagine the problem this created for scientists trying to measure the effect of the oil! They were looking for the effect caused by 1 part in 14,000 of foreign material. As an order of magnitude problem, we might ask ourselves whether other scientists studying the effects of flood sediments at a less politically sensitive time might not be tempted to overlook the paltry contribution of 1 part in 14,000 of oil.

It was soon found that the sediments partly determined the behavior of the oil. The sediments, emanating from swollen rivers, at first hung suspended in clearly defined regions. Where oil slick met silt suspension, the oil slick vanished. The explanation for this disappearing act was that the oil was rapidly adsorbed onto the surface of the silt particles, creating a physically stable mixture. The oil-catching habits of the suspended sediments prevented a potentially enormous amount of beach contamination. As it was, oil was deposited on beaches only during stormy weather and in areas where the water bore little suspended material.

As the silt particles settled, they carried their trapped oil with them. In the early weeks of the oil spill, volatile petroleum constituents in bottom samples reached a concentration of 1 percent or more in areas of maximum flood sedimentation. Then, as the months passed, the sediments were redistributed about the channel. By August, when the sediments blanketed most of the central basin, petroleum was detected in almost every bottom sample. Where sedimentation took place over an extended period, buried oil still remained in June 1970.

Near Platform A itself, bottom conditions were undeniably a mess. Analysis of a core taken on May 30, 1969, revealed an oxidized oily layer about half an inch thick, followed by a quarter inch of undegraded oil and then the expected clayey, silty, sand. There was no evidence of burrowing animals, no evidence of the usual invertebrate populations. On July 24, there were signs that the oil was diffusing into the sediments below. On October 16, the oily layer was still more diffuse, and the first signs of burrowing activity were detected. But in November there was a fresh deposit of oil on top and the burrowers had vanished once again. From that time through June, monthly tests showed that the oil was gradually diffusing and being oxidized, but still the animals did not return. Nine other stations, less severely oiled than the one nearest to Platform A, showed evidence of burrowing organisms at all times. The report did not say whether or not there were signs of unusual mortality. The interpretation of excess mortality would be difficult in any case, as reports from earlier years had already documented the extinction of bottom life after heavy floods in the Santa Barbara area.

Wherever we look in nature, there are physical and biological cycles with periods of a year, a month, or a day. We have mentioned the annual beach cycle of erosion and buildup. There is also an annual cycle of photosynthetic productivity and population density of phytoplankton. It is related to temperature, day length, rates of predatory activity, and perhaps to other oceanographic factors. Over the course of a year after the oil spill, the population density of the phytoplankton organisms varied in the normally expected way. Productivity, too, was normal, except for an anomalously low value in April 1969. (Productivity is defined as the milligrams of carbon fixed per hour per cubic meter of sample, under standard laboratory conditions. It is an approximation of the rate of photosynthesis in the sea.) The only explanation offered for the low April values was that the phytoplankton had been inhibited temporarily by large quantities of detergent applied to the oil slick during February and March.

Zooplankton populations normally fluctuate in a manner different from, but not independent of, phytoplankton. With very little background data available, it was difficult to draw any conclusions about the effect of the oil on zooplankton. Three of the stations selected for the survey were never contaminated by oil. Two of these supported the highest populations recorded in the survey—a suggestion, at least, that oil had an adverse effect on some zooplankters. Changes in the foraminifer populations, both planktonic and benthic, corresponded to seasonal cycles recorded from previous years.

Except for scattered observations related to particular research projects, there was a ten-year data gap on channel ecology. In 1959, the California State Water Quality Control Board had sponsored an oceanographic survey of the continental shelf area of southern California. This provides the only systematic study within recent years on which to base findings from the 1969 pollution study. Comprehensive as the state survey was, it was often difficult to find data comparable to those collected ten years later. How much easier research would be, if scientists could always anticipate the questions to be asked by as yet unconceived research projects!

Although the oil spill study showed that the spill per se was not disastrous, the results had an ominous portent. Since the 1959 survey, there had been a general decline in the number of plant and animal species inhabiting the channel. The most probable explanation for this is the general increase in pollution during that period. Surveys of marine flora showed a reduction in number of species at all stations. This sometimes, though not always, involved an accompanying decrease in the mass of living plants. Over the course of many years, other studies had already suggested that algal populations were declining off the coast of southern California.

One of the early signs of water pollution is a decrease in the number of plant and animal species, a phenomenon which renders the community less stable. Not only the plants are decreasing in Santa Barbara. The state survey identified 55 species of polychaetous (many-bristled) segmented worms; the pollution study found 46. Among the crustaceans, some species have vanished since 1959 and some new ones have appeared, with a net loss in species number. It is not known whether this is an example of the normal replacement of one species by another, or whether the new species are more tolerant of pollution than the lost ones were.

There is a downward trend in brittle star populations, even in areas where brittle stars have long been dominant. The dense *Listriolobus* beds, unique to Santa Barbara, are no longer so dense. Although the total area covered by the worm beds has remained about the same, the center of population density has shifted and the maximum density has decreased by more than half. At the once remarkable beds near Carpinteria, less than one tenth of the former population remains. The past success of *Listriolobus* in the Santa Barbara area may have been related to the peculiar character of the bottom sediments. It has been postulated that drilling for oil has changed the sediments during the last decade. The worms are not directly hurt by oil, however, and their beds in the vicinity of Platform A have shown less of a decline than have many of the others. So far the disappearance of *Listriolobus* remains a disturbing mystery, with nothing to implicate or exonerate either the oil or the flood sediments.

Relatively little animal mortality could definitely be traced to the oil. There was a high mortality rate only among intertidal animals that were actually coated with oil and smothered. Many kelp scallops were killed in the neighborhood of Platform A, where kelp beds trapped and held huge volumes of oil. A few mussels, sea urchins, and starfish died. Chitons, abalones, and limpets were oiled but not killed. These molluscs were frequently observed grazing on oil, and may even have helped to degrade it. The hardy anemones emerged unscathed from their oil baths. Sand fleas and beach worms escaped altogether, probably by hiding under the surface of the sand during the heavy oil dosage. About 3700 birds are known to have died, exclusive of those dying at sea, whose bodies were never recovered.

Except for the birds, only the barnacles suffered severe and long-continuing losses. In some areas, up to 90 percent of the population perished directly from suffocation. For those that survived, reproduction and development were inhibited. A study of barnacle reproduction revealed several disturbing facts. In a comparison of the number of brooding animals (those containing developing embryos) found in areas exposed to oil from natural seeps and in similar areas free from oil, it was found that there were only half as many brooding animals in seep as in nonseep areas. Among barnacles in waters polluted by the spill, the percentage bearing embryos varied inversely with the extent of oil coverage. Settlement and attachment of all species of barnacles was at least temporarily inhibited by oil.

Thus there was a large kill of mature barnacles, and this kill was accompanied by inhibition of breeding and development of the young. Many repetitions of this situation could well have serious ecological consequences. Dale Straughan points out that long-lived species are characteristic of stable marine invertebrate populations. Barnacles are a typical example. Individuals of some species do not mature sexually until they are five years old, and they may live for more than twenty years. So for them, loss of the breeding population and high mortality among the young is a very serious matter. There is a question, not only for barnacles but for many invertebrates, of whether reproduction is actually occurring in polluted areas or whether the population is being maintained by constant recolonization from unpolluted waters. There is a limit beyond which recolonization no longer can support a population. As polluted waters increase in extent and unpolluted areas shrink . . . it is a disquieting thought.

There are more insidious threats to marine life. One is the accumulation of hydrocarbons in the food chain in the same way that accumulation of pesticides occurs. These materials may be toxic or carcinogenic or both, and they would affect the animals at the top of the food chain (birds and man) most severely. *We have not yet observed this to happen,* but nature has set up an experiment for us in Santa Barbara and we have only to interpret the results. Animals in the Santa Barbara Channel have been exposed to varying degrees of natural oil pollution for much longer than people have been around to worry about it. If they are accumulating this oil to their detriment or to the detriment of their predators, someone should be able to detect the effects. Another possible effect of oil would be to serve as a vehicle to concentrate other oil-soluble pollutants like DDT in the marine environment. Together then, the oil and the pesticides or other poisons would accumulate as they climbed the food chain from prey to predator, from algae to fish to birds and man. One more hidden danger: many organisms in the sea are exquisitely sensitive to chemical messages. Chemical clues govern their reproductive, feeding, defensive, and migratory behavior. Some of the chemical messengers may be similar physiologically to petroleum constituents. The petroleum could mimic the natural messenger or inhibit the response to it. Research is needed on this important aspect of animal communication and behavior. We do know that some of the detergents used on oil slicks destroy the taste buds of fish, and thus interfere with their feeding activities.

Recolonization began almost at once in some intertidal areas, and in all, within seven weeks of the oil spill. By November 1970, most areas had a normal, as nearly as we are able to define normal, population of invertebrates. Still, the oil pollution report gives little cause for rejoicing. It is as if the oil spill were an acute attack of appendicitis and Santa Barbara, the patient. Barbara recovered quickly from her appendectomy, but the surgeon discovered that she had cancer as well. The recognition of the chronic loss of richness and variety of marine life presages ill for the coast of all southern California.

But we must reconcile the mild effects of the Santa Barbara and Torrey Canyon oil spills with the extensive damage done in Buzzards Bay. Surely part of the answer lies in the difference between crude oil and refined products. Studies of other spills of crude and fuel oil lead inevitably to the conclusion that crude is far less toxic than its refined products. Even though the crude contains all the toxic components of the refined fractions, the behavior of these components is different when they are but a part of a heavier, thicker, less soluble whole.

There is evidence that Santa Barbara crude is an unusually benign mixture, relatively insoluble and quite low in toxic compounds. In general, it must be in direct contact with an organism in order to affect it. The Santa Barbara oil spill occurred six miles offshore, and oil did not reach the shore for days. In this time, much weathering and detoxification took place. It has been suggested that bacterial degradation of the oil was unusually efficient because the oil-consuming bacterial population already existed, primed for the event by oil from natural seeps. This theory bears investigating. The question remains of whether all creatures living in the channel have acquired some degree of resistance to oil pollution, having been exposed to oil for millennia. Surely no critically sensitive organisms are living here. Studies in progress at the time of the pollution report may shed light on this intriguing puzzle. With the experiment set up by nature and the results waiting to read, no one until now has tried to get a definitive answer.

X Barristers, Solons, People, and Dons

To know when to stop
To know when you can get no further
By your own action,
This is the right beginning!
Chuang Tzu

If you looked carefully at what was happening in Santa Barbara, looked beyond the oily beaches and the dying birds and the anti-oil slogans, you could see a number of important issues raising their heads. There were legal, social, political, economic and philosophical questions to be answered, not only for Santa Barbara but for all America. The problem was so enormous that it was necessary to confront it a piece at a time; but then the partial solutions were often incompatible, and you were left once again with the discouragingly intractable whole. The perhaps irreconcilable differences between local and national interests, our dream of the "good life" for all and the environmental cost of obtaining it, our passion for exploitation and growth and our longing for peace and beauty.

The Legal Snarl

The legal questions alone are giving our laws and legal system an exhaustive test. Through the long history of the oil industry, a large body of law has grown up relating to damage caused by escaping oil. But these laws are strained to their limit to deal with the new conflicts caused by offshore drilling on federal land.

The actual origin of the pollution was the sea bed beyond the three-mile limit, which put the problem beyond the realm of local judiciary machinery. The initial effect was pollution of water, which brought into play the specific laws about water pollution. Our earliest laws were directed primarily at ships, to prevent the discharge of any material which would hinder navigation or damage boats in a harbor. As late as 1966, the law applied only to oil pollution caused by discharges from vessels; and it is questionable whether or not it would apply to discharges beyond the three-mile limit, on the high seas.

The Outer Continental Shelf Lands Act of 1953 requires that federal lessees conduct their offshore operations according to the regulations established by the Interior Department. This mention of the operating regulations by the law includes them by reference in the law. As the regulations can be changed at the discretion of the Interior Department and still have legal status, management of offshore operations should be less cumbersome than it would be if every small procedural change required legal action. Unfortunately, the regulations in effect at the time of the Santa Barbara oil spill merely exhorted the lessee to be careful, without specifying what was to be done or who was to be liable in the event of a pollution incident. Secretary Hickel recognized this flaw in the regulations at once and quickly amended it. After February 21, 1969, the lessee was to be held fully responsible for any pollution damages caused by development of the lease. This did not solve the problem for Santa Barbara. It raised the issue, instead, of whether or not the new regulations could apply to the situation which had inspired them. The state of California and the county of Santa Barbara, their governing bodies sensitively tuned to the local needs and fears, had numerous laws which were very explicit about liabilities for water pollution. The applicability of these was a difficult question for the courts because although the pollution originally occurred on federal land, its main effects were on the hapless state of California and Santa Barbara County.

It did not take long for the first suits and claims to be filed. On February 18, 1969, the State of California, the City and County of Santa Barbara, and the City of Carpinteria together submitted a claim to the Secretary of the Interior. This Claim for Damage or Injury, Standard Form 95, is the same one you would submit if you stumbled and sprained your ankle on the steps of a federal courthouse. The "slip and fall" claims are usually paid off without a murmur. California's claim was a different matter. The basis of the claim was that the claimants' Fifth Amendment rights had been violated. That, through the negligence of Interior Department officials, they had been deprived of the use of their property, without due process of the law. The amount of the claim was half a billion dollars.

There are some indications that the Department of the Interior was not as watchful of its profit-seeking charges as it should have been. Former Secretary of the Interior Udall not only publicly shouldered the blame for the decision to allow drilling in the Santa Barbara Channel, but many of his statements indicate that he suffered a great deal of private torment over the decision, both before and after the oil spill. He admitted that, although the geology of the channel was known to be unusual, the question of tighter drilling regulations had never been seriously considered. Perhaps the oil and gas which escaped as Platform A was being installed should have been taken as an omen. Perhaps the casing program submitted by the

Union Oil Company and approved by the Regional Oil and Gas Supervisor was inadequate. Perhaps there should have been inspections of drilling operations before the well was completed, although it is difficult to imagine how any inspection could have revealed that the drilling team was going to withdraw the drill pipe too fast and, in general, fail to respond appropriately after the kick began. Probably a more meaningful question is, if certain government officials *were* negligent in their duties, was this the cause of the blowout? Would stricter regulations and a more hawkish scrutiny of operating procedures have prevented it? The answer appears to be no. So although California's claim highlighted the fact that in the past the Interior Department has been a bit casual in its supervision of the oil industry, the claim, or others like it, had small chance of success. For one thing, as our system now works, it is most unlikely that any court would render a decision holding a government regulatory agency responsible for the acts of the industry it regulates, when the issue was merely one of surveillance. This would be an unfair and unbearable burden on the government, and a sign for industry to get away with as much as possible without getting caught.

California, Santa Barbara City and County, and the city of Carpinteria also filed suit in the California Superior Court against the Union Oil Company and its partners, the Peter Bawden Drilling Company, and various concerned agencies in the federal government. This suit was a class action on behalf of the plaintiffs themselves and of all other "public entities and agencies" of the state that had suffered damages. The amount of this claim was $560 million. Union Oil responded by filing a crosscomplaint in which it charged that the driller's negligence had caused the blowout. The oil company sought not only a judgment holding the drilling company responsible for damages claimed by the plaintiffs but also reimbursement to itself for some $25 million spent on cleaning up after the spill. This suit raised both jurisdictional questions and questions of which laws were applicable. An absolutely strict interpretation of existing law would apply one body of law to the Outer Continental Shelf, another in the territorial waters and shore, and still another on the high seas.

Another class action, a private one on behalf of citizens of Santa Barbara, was filed, claiming a total of $1.3 billion in damages. Among the defendants named were Walter Hickel, William Pecora, Don Solanas, John Steinhart, Lee DuBridge, and the Union Oil Company and its partners. These three suits alone amounted to claims exceeding two and a third billion dollars. But who can measure the cost of an oil spill? How much is a bird worth, or a lazy afternoon at the beach? Certainly the actual economic loss—of property and to the fishing industry and local business—was only a fraction of that claimed. As we have seen, the ecological damage was not severe. Much of that astonishing figure of more than two billion dollars must then be laid to the cost in human anger and grief, if not to a desire to punish those who had wronged Santa Barbara.

The newspapers hopefully reported each legal action as it came up and sadly listed the cases that had been thrown out of court for one technical reason or another. We are reminded of Charles Dickens's novel *Bleak House*, in which most of the characters were litigants in Chancery Court. None seemed to understand what points were actually at issue in their cases, which droned on for years until the attorneys had used up whatever funds were available. The cynical message of *Bleak House* is relevant today, when hurt and angry people turn to the courts for redress of grievances and find instead, as Santa Barbarans found, a legal maze of jurisdictional squabbles and a group of suits that seem largely political in intent.

The litigation surrounding the blowout offers opportunities to make some crucial judgments in American law. Sooner or later these judgments must be made, even if the challenge of Santa Barbara is passed by through cases being settled out of court or on technicalities in court. At least three basic issues are open: What obligation has a polluter to pay for damages caused by a pollution source under his control, even if he did not willfully or negligently permit the pollution to occur? What rights, if any, does a citizen have to an unpolluted environment? And who owns the sea bed and which laws apply there?

Though not the specific subject matter of the law suits, there are some other fundamental issues of society involved. They include the appropriate balance between local and national control, the rights and responsibilities of the federal government as trustee for public lands, and the mechanisms available for citizens to seek participation in decisions affecting their future. David Walmsley of the Harvard Law School has prepared a study of the legal implications of the oil spill which gives cause for reflection.[1] It seems possible that our legal system, along with many other of our cherished institutions, may be unable to cope with the complexity of our society's problems.

There were other suits and countersuits. The American Civil Liberties Union asked for a federal court order barring the federal government from implementing any of the recommendations in the report of the second panel. Other groups asked that drilling be banned. And then, on April 10, the Pauley Petroleum Company and five other oil companies sued the federal government for $230 million. They claimed that the new drilling regulations made development of their leases economically and practically impossible, and they wanted reimbursement to cover the amount they had paid for their leases plus their expenses and losses of known oil revenue. One spokesman for Pauley stated that there was $230 million worth of oil in the leases. Presumably, the government was to be glad that the suit was not larger than it was. It was rumored by those close to the development of the channel leases that Pauley had found its lease parcel to be less valuable than it had guessed when it bid for the tract. This suit was perhaps a way to get out of what had turned out to be a bad deal.

The Lawmakers

While the legal apparatus of the nation began with glacial slowness to move toward the issues raised in Santa Barbara, members of the legislative branch of the government had been active with public and private efforts to make themselves heard. For months after the blowout, the *Congressional Record* was filled with speeches, statements, and insertions "for the record" of letters, newspaper accounts, and comments of all kinds. The brave men who fought the spill were eulogized; the birds were eulogized. Ecological doom was promised unless drilling was halted; a no-energy doom was promised if industry was tampered with. Gladwyn Hill's emotional outpourings in the *New York Times* are enshrined in the *Congressional Record* almost alongside the dispassionate analysis of the *Washington Post's* Spencer Rich.

Numerous bills referring more or less directly to Santa Barbara were introduced into both the Senate and the House of Representatives. At least six such bills were introduced in the Senate and a larger number came into the House. A look at the proposed legislation in the Senate will illustrate the situation.

Four bills, S. 3093, S. 1219, S. 2516, and S. 3351, called for an end to drilling in the channel, suspension of production, and studies of present leases and means for terminating or exchanging them. These bills bore the names of the senators from California, Cranston and Murphy, who were joined by the environmentally active pair of Edmund Muskie and Gaylord Nelson. One bill, S. 4017, presented the administrations's point of view. Another, S. 1592, was introduced by Senators Edward Brooke and Ted Kennedy of Massachusetts, perhaps to take advantage of the concern for Santa Barbara in order to protect areas off Cape Cod where oil drilling had been considered from time to time.

These bills had only the remotest chance of becoming law. Nor did their authors expect that they would ever come to a vote before Congress. Most bills introduced into the Congress are for other purposes, a phenomenon which many Santa Barbarans probably did not understand, as their hopes flew up and down like yo-yos with the daily reports of Congressional activity. Some bills are mainly for public display in the author's home district. Some are introduced in order to promote hearings, often in an effort to provide a forum for expression of opinion or to expose information. Sister bills with slightly different emphasis may be introduced to provoke hearings before still another committee which may be more sympathetic, or of which the author is a member.

The rivalry between Senators Jackson and Muskie in the field of water pollution legislation was well underway before the oil spill. The senators continued their competition over the Santa Barbara affair. Most of the Santa Barbara bills were referred to Senator Jackson's Committee on Interior and Insular Affairs, which held hearings in Santa Barbara in March 1970 and in Washington, D.C., in July 1970. These hearings served many functions for the committee members and for the witnesses, but the overall tenor was more political than technical, and more for purposes of show than for legislative change. It is striking that the hearings of March heard 26 witnesses with federal, state, or local political affiliation, 17 private citizens or representatives of citizens' or conservation organizations, *three* technical experts, and *one* industry representative. The second set of hearings brought out a few more men from industry, but the general situation was the same. If there had been serious intent to enact legislation, far more attention would have been directed to industrial problems and the technical facts.

There was a clue in the bills themselves that the ouster of oil companies from the channel was not expected. Termination of the leases would have necessitated return of the $603 million paid for the leases, in addition to some amount of reimbursement to the oil companies for their expenses. But the bills under discussion contained no appropriations for this purpose. The behavior of the oil industry toward the clutch of hastily written bills gives an insight into the way Congress functions. Industry testified when summoned, but without bringing out its big guns. In the cloakrooms, however, and the private offices and the Metropolitan Club, the oil industry was making its case through the most influential lobby in Washington. Friends of oil in Congress promised that nothing hasty would be done.

As a contingency, the White House assessed the possibility of forcing the existing leaseholders out of the channel. An internal memorandum prepared by the Office of Science and Technology revealed that the cost of repaying the purchase price of the leases plus legitimate expenses and other oil company claims would total more than a billion dollars. If court judgments turned out to be unfavorable, the costs could have mounted to several billion dollars. Since the bills in Congress contained no appropriations of this nature, the White House could only conclude that removing the oil companies was neither the intent of the bills nor even a possibility. Congressional action was needed to appropriate such monies. The administration did nothing to encourage the appropriation.

One other possibility remained, one suggested by California's Representative Charles Teague and others. Require the oil companies to trade their channel leases for leases in the Navy's Elk Hills petroleum reserve or in other federal lands. This would circumvent the problem of actual cash outlay, but it presents, in the case of Elk Hills, other difficult problems. The oil reserves of Elk Hills are proven, and the potential of most of the channel leases was (and still is) unknown. In a trade of something of unknown value for something of fairly clear value, one suspects that industry would either get the better of the deal or go to court.

Local Action

Regardless of the national flurry it aroused, the story of the Santa Barbara oil spill is basically a story of the people of Santa Barbara and their struggle for the right of self-determination. There are contradictions and inconsistencies in their opposition to the oil industry. "If the oil had been spilled forty miles down the coast," said one Santa Barbaran, "it would have been just another gummy day in Ventura, and nobody would have much cared. But it hit *here*. . . ." But it hit here. Where the per capita holdings of oil company stocks were about as high as in any American city. Where, in order to carry through threats of boycotting all service stations connected with offshore leaseholders, Santa Barbara would have earned the distinction of being the first American city to abandon the automobile. Where the city supported Stearns Wharf as a service facility for oil operations on state leases.

Except for being sorry that the spill had occurred, Santa Barbarans were not united in their feelings about oil. Most of the citizens were among the renowned Silent Majority, who either did not express their opinions or did not have any. Probably everyone not directly connected with the oil industry in Santa Barbara would *prefer* that there was no oil production in the channel. Most people were willing to sign petitions to that effect. But it was still a minority that was ready to do more than just wish that oil weren't there. A few clear voices even managed to break through the anti-oil stand of the Santa Barbara *News-Press* (called the *News*-Sup-*Press* by some) and point out, in letters to the editor, the invalidity of the radical position. The Channel City Club and its members tried very hard to obtain a balanced view of the problem, without being swept away in the emotional orgy that gripped the more radical citizens. Many businessmen simply tried to minimize their losses, and they too were divided over the best course of action. When the Chamber of Commerce accepted funds from the Union Oil Company in order to promote advertising of the still extant joys of Santa Barbara, many people grumbled that there was only one name for someone who cries rape! and then accepts money from her ravisher. One suspects that much of the activity was for show, and was carefully calculated. Was it better to minimize the damage that had been done, or would it be more profitable in the long run to claim crippling injury and hope to collect outrageous amounts in damages? Were the piles of oil-soaked straw from cleanup operations kept around longer than necessary, fire hazard though they were, in order to keep alive the protests? There were those in Santa Barbara who did not want the mess cleaned up quickly, because that would emasculate their anti-oil campaign.

On January 30, 1969, the organization GOO! was born. The story has been told many times of how artist Bud Bottoms, long a foe of oil, exclaimed to his boss "We've got to get oil out!" The boss was Marvin Stuart, a General Electric public relations man with the necessary ear for slogans. As the story goes, he cried out "That's it! GOO! We'll call it GOO, for Get Oil Out." With relish, Bottoms and Stuart made plans for a mass organization named GOO, enlisting the help of former state senator Alvin Weingand, who became GOO's leader. There were ideas and offers of help from all sides. The bus company provided free rides into town on February 1, when all residents were urged to walk, bike, or bus to the city parking lot at Cabrillo Boulevard and Santa Barbara Street to stage a mass burning of their oil company credit cards. February 3 was designated "Black Monday." Taking a lesson from the college generation, the staid residents of Santa Barbara held protest rallies and demonstrations. "Ban the Blob." "Lose the Ooze." They got plenty of publicity but, to their profound frustration, little else in the way of satisfaction.

Other action groups were formed. Santa Barbara Citizens for Environmental Defense, S.B.C.E.D., chose the legal route for its main thrust. It supported various suits with money, by finding expert witnesses, and in enlisting the services of the American Civil Liberties Union attorneys.

Of all the things that the citizens of Santa Barbara did, it was GOO that captured the public's sympathy and imagination most completely. Perhaps it was only because of Stuart's genius in giving the organization its name, but probably it was more than that. Probably it was a genuine upwelling of grass roots sympathy. Thousands of people all over the country and from as far away as Australia signed GOO petitions and wrote letters of consolation and shared grief. Citizens of San Pedro, California, founded a sister organization called GOO TWO. The editors and staff of the *News-Press* were unashamedly on the side of GOO from the moment of its inception. Without the press, GOO's campaigns could never have attracted the attention that they did.

From the beginning, there was no question about where the *News-Press* stood on the oil issue. It may have crossed over the line between a newspaper's traditional role of reporter and the hazardous role of advocate. It went out of its way to be helpful to out-of-town newsmen and, some people thought, to keep the Santa Barbara story alive in national newspapers. There is no doubt that the lurid accounts of the oil spill sold a lot of newspapers. So many, in fact, that reporter Robert Sollen was assigned full time to cover oil pollution, and there were numerous contributions by other writers as well.

News media far from the scene were guilty of questionable reporting. In testifying before a Senate subcommittee, Fred Hartley, president of the Union Oil Company, said something to the effect that, considering the total size of the accident and scope of its reverberations, he thought an undue emphasis was being placed on the death of birds. The *New York Times* misquoted him, out of context to boot: "I'm amazed at the publicity for the loss of a few birds." The *Wall Street Journal* quoted the *Times*. The *Washington Post* did, too, and angrily editorialized about it. On NBC, David Brinkley spread the story all over America. But Hartley had not said that, not at all! He was no tenderhearted bird lover, but neither was he stupid. And no amount of retracting by the newspapers or advertising by Union could erase the impact of that callous misquote.

And the University

It is interesting that the students at the University of California at Santa Barbara did not enthusiastically embrace the local cause. The radical center of protesters saw the blowout as just another manifestation of Korporate Amerika, and the menial work of town protest and cleanup was not allowed to stand in the way of the real work of the revolution, or of ending the war in Vietnam. Student volunteers did enlist in many programs. Ecology action groups joined or worked in parallel with GOO and other groups, to try to halt oil production. But there was nothing like the strong feeling of unity that has moved so many campuses in times of crisis. Most of the students and most of their professors, like most of the townspeople, felt very sorry and more than a little angry. Many performed some token ritual in recognition of the tragedy before settling back to the day-to-day pressures of their own lives.

Modern public universities are part of the "Establishment." Depending as they do on state appropriations and federal grants, they not only are state agencies by law, but they must act "responsibly" if their financial support is not to be jeopardized. It is pointless to expect the administrators of a *public* university to take vigorous public stands on political issues, whatever they may feel in private. University presidents and chancellors who take positive stands usually join the growing ranks of ex-presidents and ex-chancellors.

Gordon MacDonald, U.C.S.B.'s Vice-Chancellor for Research, served on both of the presidential advisory panels. In the end, he concurred with all the recommendations, although with occasional reservations. But MacDonald, as former vice-president of the Institute for Defense Analysis and with a long history of involvement in high level government advisory work, was as much a part of the current inner circle of government as any Washington bureaucrat.

U.C.S.B. acted as host to the first meeting of the President's Panel on Oil Spills. During the meeting, a group of faculty members not associated with the panel issued news releases and held a news conference of their own. The positions taken were predominantly anti-oil and anti-government. These people were speaking for themselves, in the tradition of the university. The university took no stand.

The oil spill activists among the faculty were all young. At times they ventured dangerously close to the line that divides scientific fact from concerned opinion. They were eagerly embraced by citizens' movements and exhibited as the experts who knew how things really were. But as we have seen, nobody knew how things really were. Either the information had not yet been made public, or the data had not yet been obtained. It is clear that the young scientists captured the public's ear because of their supposed standing as scientific experts. What is more, they were continually urged to express opinions well beyond their personal expertise. It is not surprising that a few succumbed to the seduction. So a geographer talked knowingly about geology (the two fields have some common knowledge, but not much); a marine biologist talked a great deal about birds and economics; and a geophysicist rendered judgments about petroleum engineering. The situation appeared even worse than it was, for the quotations and public reportage invariably left out the qualifiers and uncertainties in which these men couched their words. The geophysicist left U.C.S.B. shortly after the oil spill. The circumstances surrounding his departure have not been made public, but it is probable that his activist role hastened his departure. It is notable, even astonishing, that the older and more famous men associated with environmental problems at U.C.S.B. played no visible role. Should they have? What kind?

The Predicament

Many people maintain that we cannot put a price tag on beauty. The answer to that is, we must. Somehow, we must order our priorities so that we can, rationally, answer questions like "Is it more valuable to the nation to have the oil and oil revenue from Santa Barbara, or to keep Santa Barbara unspoiled by industry for her residents and visitors to enjoy?" To most Santa Barbarans, the answer is absolutely clear and there is no room left for discussion. Unfortunately, the answer is also absolutely clear to some other groups. And the two absolutely clear answers happen to be diametrically opposed. There are compelling arguments on both sides. A compromise would seem in order. The Santa Barbara County Report of November 1967 seems like a fair and reasonable compromise. But after the blowout, polarization of the two sides became so extreme that, as usually happens when emotions run high, meaningful discourse became impossible.

The Santa Barbarans made the mistake of trying to argue their case on technical and ecological grounds. On both counts, their arguments can be shot down. In most of the channel, drilling for oil is less hazardous than in the Gulf of Mexico. If danger is the password to freedom from oil wells, the Gulf should be excused before Santa Barbara. As for the ecological argument—we have seen that the channel has been naturally polluted by oil for centuries, and that relatively little ecological damage could be laid to that particular oil spill. The threshold of sensitivity to oil of the organisms living in the channel may in fact be unusually high. If this is so, oil production in the channel should proceed at full speed and we should spare areas as yet unexposed to oil. What seems to have happened is that the Santa Barbarans confused ecology with beauty and recreation. When applied specifically to Santa Barbara, the ecology issue is a red herring. The other point, that the channel is loved for its beauty and enjoyed by pleasure-seekers, *is* a valid one and one that will have to be reckoned with in the case of Santa Barbara and in all similar cases yet to come.

At heart, it is a problem of the majority versus a minority and the difficulty of allowing the majority to rule without oppressing a minority. One way of asking it is, what rights have the Santa Barbarans (or any citizens) to live in an unpolluted environment? But another way is, what rights have the Santa Barbarans to tell the entire nation, represented by the federal government, how it may use its public (belonging to everyone) lands? This, even overlooking the fact that the record of the city of Santa Barbara and the state of California was far from clear as far as encouraging the oil industry was concerned.

We cannot have it all ways at once. How much are we as a nation willing and prepared to sacrifice for the enjoyment of beauty and a healthful and unpolluted environment? It will have to be a great deal, perhaps more than we are willing to sacrifice. How hard are we willing to work, as communities, for the right of self-determination? It will have to be very hard. It is easier to grumble and protest than to take a firm and constructive hold on our own destiny.

XI Energy

Men at some time are masters of their fates:
The fault, dear Brutus, is not in our stars,
But in ourselves, that we are underlings.
William Shakespeare

Man's cultural history includes about the last 10,000 years. For 9,000 of those 10,000 years human energy requirements changed but little: about 2,000 calories a day in food, supplemented by animal power, crude water power, and—directly or indirectly—solar energy. The supplemental energy used by each individual was about equal to that used by one 100 watt light bulb. The total energy requirement can be understood by noting that the world of 2,000 years ago housed only about 150,000,000 people. At present, we add the population of that Roman world to our world about every two years.

Fossil Fuels: A Historical Romance

After those most primitive times, there was a long, slow increase in energy use to a daily rate of about 500 watts per person during the period before the Industrial Revolution. This additional energy was supplied by occasional use of water power, animal power, wood, and, beginning about 800 years ago, coal. About 150 years ago our energy requirements per person in the Western World (and especially in the affluent portion of it) began to escalate to the present average in the United States of about 10,000 watts per person. During this period, most of the energy has been supplied by fossil fuels. These fuels were deposited over the course of millions of years. The time needed to replenish them may be of the order of 100 million years or more. For practical purposes they are a one-time supply, and cannot be replenished at all. Today, petroleum is dominant among the fossil fuels. Between 60 and 70 percent of our industrial energy requirements are filled by petroleum and natural gas, which are often found and produced together.

The astonishing growth of the past cannot continue. We have estimates from all sides, from radical ecologists to representatives of industry, but they all carry the same message. We cannot go on much longer at this rate with our primary reliance on fossil fuels. Over the total span of human history, what seems normal to us in our lifetime will appear as a peculiar aberration, in which man began a massive and sudden exploitation of the fossil fuels stored within the earth and in a very short time exhausted them. M. King Hubbert offers a detailed analysis showing that the peak of possible petroleum production already has been reached in the United States and that regardless of which estimates are used the peak in world production will occur within the next 30 years. If these estimates are wrong by a factor of two, it makes little difference even in the short run.

There are only two ways out of this growing shortage in fuel. The first is to reduce our demands for energy and the second is to find other sources. The other source most commonly proposed is energy from nuclear reactions, either fission or fusion. Although experts expect the breeder reactor to become a reality so that our supply of fissionable fuel can be made to last a very long time, we *presently* have no successful breeder in operation that can meet the demands for expanded energy supply. Furthermore, we have been having second thoughts about the possible side effects of a heavy reliance on nuclear fission for our energy supply. The outcome is uncertain, because we cannot be sure until we have it that an adequate breeder reactor can actually be built, nor can we know whether the side effects impose penalties too severe for us to bear. For fusion, the situation is even murkier. Controlled fusion is not yet possible, and to some experts the difficulties seem insurmountable (although other experts are confident that all obstacles can be overcome). Before swallowing the claims that fusion is a safe, clean process with no undesirable side effects, we should recall that in the 1940's similar claims were made about the cheapness, safety, and ultimate harmlessness of nuclear fission. We will learn more about the penalties for controlled fusion if and when controlled fusion becomes a reality. At the moment we have no guaranteed option for the future energy supplies that we must have if we intend to continue our present way of life. In the long run, we will rely less on fossil fuels and either reduce our energy consumption or find another source. There is no other possibility.

If we accept the principle, upon which scientists and most of our society depend, that the past is a necessary guide to predicting the future, we must note that each innovation of the past (whether it be fertilizers, DDT, tobacco, digital computers, the corporation, the automobile, or money) has carried with it ugly, unhealthy, or dehumanizing side effects that we did not foresee or that were larger or more damaging than anticipated. This "Law of the Negative Bonus" has the same universality as has any law in physics. We ignore it at our peril. We should also note that the degree to which we foresee the difficulties attending a proposed innovation varies inversely as the distance (in time, engineering understanding, or both) we are from actually executing the innovation. So far it has not been possible to place any upper bounds on the limits to which these difficulties may go. So although we cannot tell how great the problems associated with nuclear energy may become, we can be certain that they will be larger than we expect.

Thus far we have neglected the fact that very large portions of the world still exist at early, low-energy levels of civilization. In the United States we mouth sentiments that everyone in the world should be able to share the material benefits that we have fought so hard for or taken so ruthlessly. Yet we know that if people were to try to do so, in the case of energy, at least, we could not meet the demand. There is a discrepancy here. In the light of our current policies, it seems that we don't really intend to have social justice.

Since World War II the gap between the rich nations and the poor nations of the world has grown larger. Six percent of the world's population in the United States consumes more than a third of the world's production of energy. The large majority of the world population being shortchanged may one day take offense and fight for a larger share of energy and other resources. Thus, whether from humane considerations or merely in a self-defensive hope of preserving the world from intolerable conflict, both our actions and our policies must change. Against this backdrop, the Santa Barbara story assumes a different kind of meaning.

The U.S. Energy Crisis

Our recent experiences with brownouts, blackouts, and "temporary" shortages in natural gas supply are viewed by most individuals as an annoyance, or as the failure of a self- and profit-centered industry to provide service, or even as the failure of an all-protective government to protect. But M. A. Wright, Chairman of the Board of the Humble Oil and Refining Company, has written an article about the "U.S. Energy Crisis."[1] He says, for example:

> In absolute terms, future energy demand can be described in a single sentence: The outlook is for a near doubling of total U.S. energy demand between now and 1985. . . . Domestic natural gas supply is no longer able to keep up with the rapid growth in demand for gas. . . . Nuclear energy supply is running behind earlier expectations, but appears to be on the verge of rapid growth, *provided* environmental and siting difficulties are overcome. . . . The ability of coal to contribute to energy supply today has been limited by a shortage of economic spare production capacity, in part a consequence of *over-optimistic nuclear forecasts, by local air quality regulations* restricting the use of high-sulphur coals, by *recent mine safety regulations* and hopper car shortages. . . . Present and future supply of hydropower is limited by availability of economic dam sites. . . . By 1985 *imports are expected to supply more than 60 percent of U.S. demand for petroleum.* (Italics ours.)

The crisis is real enough. Day by day and season by season, arrangements to supply oil and gas to individual and industrial consumers grow more ingenious and complex. Pipeline diversions, new interconnections in electric power grids, management of fuel storage facilities, have reached a level of subtlety and complexity undreamed of in the everyday world of all but a few. And yet, in those vanishing places where supplies and capacities still exceed demand, we are exhorted to "live better electrically" or "use gas, the clean fuel," or otherwise to increase our consumption of energy.

Notable in Wright's article is the absence of any mention of how we might accommodate, in whole or in part, our stated goals of helping those in low-income countries to reach a comparable level of energy usage. Nor should we expect such considerations to enter from this quarter. We are told that oil imports must supply 60 percent of our requirements in 1985. What if this should be incompatible with the needs and aspirations of the nations from whom we propose to obtain this petroleum? For many years, most of our imported oil has been produced by American corporations abroad. Governments rise and fall, negotiations go on, and Western Europe, Japan, and the U.S.S.R. all compete for foreign oil. Of course, some capital flows into the poorer countries as a return for their oil concessions. But by any fair accounting it has been a bad bargain for them. The oil is taken away, a large part of the profit is taken away, and they are left only with such small receipts as they could negotiate. Frequently the search for foreign oil has resulted in our government's supporting unpopular, authoritarian regimes or military dictatorships in order to "protect U.S. interests"–which often means, oil. The poorer nations are beginning to understand that they are being had.

To meliorate the current energy problems, Wright proposes expansion of nuclear energy supplies through technological advance and development of plant siting procedures. He wants to encourage offshore exploration and accelerate offshore lease sales. He recommends price improvements for oil and gas, reduction of regulatory obstacles, and additional funding for research and development. But what to Wright are stubborn obstacles to the development of energy resources result from attempts to cope with harmful side effects, from our government's attempts to protect our interests, and from the inability of the public purse to solve all technological problems simultaneously.

The Economics of Oil

An argument frequently used by those wishing to get oil out of the Santa Barbara Channel was that the artificially high price of oil in the United States was the only reason it was profitable to exploit the oil there. If we had a free market in oil the price would fall below the level at which oil could be produced profitably in the channel. To explore this question we must examine briefly the economics of oil.

We do not have anything like a free market economy in any of the fuels industries. Probably we never did have, although the turbulent days of the late 19th century industrial barons are sometimes regarded in that light. Seven major corporations dominate the world petroleum market, controlling more than 70 percent in 1969. The Union Oil Company is one of the largest of the many smaller companies that, together, control another 15 percent of the market. Somewhat less than 15 percent is controlled by nationalized oil companies. Such an industry is called oligopolistic, or occasionally monopolistic, in its structure. An account of the economic structure of such industries is given by Chamberlain in *Theory of Monopolistic Competition.*[2] Monopolistic groupings are not unusual. The steel industry is a similar grouping. When such an industry has essentially undifferentiable products, as have oil and steel, they inevitably engage in price-fixing, sometimes by actual collusion and sometimes by custom, following one another's lead in price changes. The sensitivity of each firm to each of the others could hardly have any other result.

Such industries also become politically sensitive and they interact extensively with the government. At best, this is a search of a wise businessman for ways to a better and more profitable future for himself and the country. At worst it is greed and self-interest made powerful. The existence of these industrial groupings guarantees that power and government influence depend on the ethics of a great many men who are subject to no public check.

A monopolistic industry customarily finds itself enmeshed in government regulation. Such is certainly the case with oil. Here and abroad, the search for oil and gas, leasing, production, and a great many technical details are subject to at least nominal government regulation, although it sometimes becomes a question of who is regulating whom. The retail sales of petroleum products at service stations constitute between 7.5 and 8.0 percent of all retail sales in the United States. The oil industry and the government cannot afford to ignore each other.

There is a sorry history of oil's political influence in Congress. Men like Lyndon Johnson, Treasury Secretary John Connally, the late Sam Rayburn and the late Senator Robert Kerr (of Kerr-McGee Oil) have been vigorous proponents and defenders of oil's special regulatory and favored tax situation. However, not only the federal government is concerned with oil. The regulatory bodies most directly concerned are the various state groups linked together through the Interstate Oil Compact Commission. This organization, established in 1935, systematized the control of production. It covers total production within each state, spacing of wells, and general production practices. The total quotas are specified for each oil field and frequently for each individual well.

Under such controls the economics of pricing is very strange. Production quotas are derived from estimates of major oil companies of what they expect to require from each state, and from monthly estimates of demand prepared by the U.S. Bureau of Mines. Prices are assumed to remain constant or to rise slightly. But if output is set to meet some expected demand at some expected price, a necessary outcome is that the assumed price will turn out to be the actual price. Thus the control of production determines the price of oil.

Oil's tax benefits also permit the production of oil in areas that might not be profitable without the tax breaks. Most U.S. industries pay corporate taxes amounting to between 40 and 50 percent of their total profits. Oil companies pay between 1 and 10 percent, and some pay nothing at all. Oil companies may deduct 23 percent of their income as a "depletion allowance." The origin of this situation is rooted in history, involving compensation for capital depletion, promotion of a more competitive market, and other factors all of which now appear largely irrelevant. Much has been written about the inequity of the depletion allowance and many attempts have been made to reduce or eliminate it. Granting a depletion allowance is tantamount to rewarding the oil companies for removing a depletable resource which we wish to conserve. In addition, tax incentives distort the market price, denying the user the choice of other competing fuels and maintaining artificially high demand.

Additional tax breaks are afforded the oil industry by permitting liberal write-off of capital expenses and certain indirect costs as current production costs. Other special benefits include permission to deduct certain foreign operating expenses whether or not that operation showed a profit.

For the now legendary lone "wildcatter," there may have been some need for help in a risky business. For today's monopolistic industry grouping, however, the risks are well-known, anticipated, and shared. There is no good reason for the continued favored tax position of oil. We need to find out what is the least disruptive way to change the system, and then to change it. But oil's influence is strong. Change may be a long time in coming.

Oil Import Quotas

The United States has been a net oil importer for the last 25 years. Tariffs on imported oil have existed since the 1930's, but the competition from cheaper foreign crude oil became of concern only in the 1950's. Import quotas were at first voluntary, but in 1959 they were made mandatory at 1957 levels. They have been modified several times since then.

The stated reason for oil import quotas is for "national security." The idea is that, in the event of a war, the United States must be self-sufficient in oil, and unrestricted imports would mean a reduction of U.S. oil production capacity below defense requirements. It is the Office of Emergency Preparedness (formerly the Office of Emergency Planning), working with the Department of Defense, that is responsible for estimating defense requirements. As recently as 1969, the scenario projected from which to plan for oil needs (and also for 91 stockpiled commodities from platinum to feathers) was: how much would be needed if the United States were engaged in two World War II's simultaneously, with only a 10 to 20 percent reduction in normal peacetime usage? The origin of this idea is unclear, but today it seems absurd. President Nixon has recently permitted reduction of some stockpiles of metals, in recognition of this fact. But with the oil lobby remaining the most powerful in Washington, it will be difficult to end import quotas in the face of determined opposition from industry.

The free market world price for crude oil is about $2.00 per barrel, at a time when domestic prices are about $3.00. From this it is possible to estimate the magnitude of the cost of this particular bit of "national security" we think we are maintaining through oil import quotas. Although it is difficult to estimate what share of the U.S. market imports would take if restrictions were removed, we have a clue. Residual oils (what remains after light fractions have been removed) used in this country consist of 50 to 60 percent imports, compared to 20 percent or less of imported crude oil. Quotas are much less restrictive against residual oils. If corresponding shares of other petroleum fractions were imported, total costs to consumers would be from one to four billion dollars per year less than they are now. Thus several billion dollars are absorbed from public and personal income, making the assumptions behind oil import quotas some of the most expensive assumptions in our present defense costs.

There would be other effects of changing the import quota policy. From the U.S. point of view, balance of payments in international trade would worsen, at least for a time. Much offshore production, including that in Santa Barbara, would probably be unable to compete in the lowered price market. Economy of operation in the domestic oil industry would receive a stronger stimulus. But although the situation is complex, it is time for a serious look at alternatives to our present policies, especially in the light of the limited total supplies and the possible trade-offs with other fuels.

Alaskan Oil: Another Santa Barbara?

The discovery of vast oil deposits on the North Slope of Alaska occasioned glowing reports in the newspapers of the country. This oil, held out as our hope for energy in the immediate future, has also stirred bitter controversy. The dispute over construction of the Alaska pipeline has brought an extended confrontation between conservation groups and economic developers. Two basic questions must be resolved. The first, are we going to produce the oil on the North Slope, is the kind of question that in the past has been asked too infrequently in this country. If we find a source of something useful, *of course* we will exploit it because, like the mountain, it is there. Fortunately, there are some who challenge this approach to our environment, although the ultimate goal of industry and our economic system still seems to be the uncompromising rape of the earth. The second question, assuming that development of the Alaskan oil fields is inevitable, is how will we get the petroleum to market.

Oil from the North Slope could be taken to market by a combination of pipeline to southern Alaska and then by tanker, or by tanker all the way. The voyage of the super tanker Manhattan through the Northwest Passage showed that tankers can negotiate the trip. But given the comparative spill rates from tanker transport and from pipelines, pipelines seem by far the safer means. The decision in favor of a pipeline is bolstered by our ability to contain a spill on land, compared with our helplessness to control a spill at sea.

Spurred on in part by the voyage of the Manhattan, Canada passed an Arctic Waters Pollution Act. In this act, she asserts her authority over the waters contiguous to the Canadian arctic islands. She declares her control of the Northwest Passage and makes no provision for "right of innocent passage." It is conceivable that, under this act, the Northwest Passage may be closed to tanker traffic. If so, the pipeline will be the only possibility.

There are technical solutions to most of the specific objections to the pipeline, although as in the case of Santa Barbara, the basic objection may be that some people just don't want it there. But this does not mean that all the problems can be solved easily, or that deleterious environmental side effects can be prevented. Our "Law of the Negative Bonus" guarantees otherwise.

Offshore Oil

Offshore oil production is a rapidly growing segment of the oil industry. By 1968, one seventh of all domestic oil production was offshore. One third of production is expected to be offshore by 1980, and some estimates say the figure will be closer to two thirds. Income to the federal government for lease sales and royalties has been averaging about $1.5 billion per year. The states have about three quarters as much production on their leases, and they derive considerable revenue from such production. Private investment in the 2500 offshore structures on both state and federal lands exceeds $10 billion.

In California's offshore lands, slightly over 2 percent of the production is from federal leases. The remaining 98 percent (almost five billion gallons in 1968) is from state leases. This makes California the leading producer of offshore oil on state leases, by a considerable margin, although if present plans are executed Alaska may soon take the lead. It was offshore development that enabled the California oil industry to stage a remarkable comeback after a 24-year low in 1960. At the time of the blowout, one out of four California oil wells was offshore. There is a certain logic in this pattern, because Californians are the leading consumers of gasoline, both on a per capita and on a total consumption basis.

Concomitant with the pressure for expansion of domestic offshore production is the increasing difficulty of getting foreign concessions. Algeria, Indonesia, Nigeria, and others have claimed larger shares in oil revenue. Peru has expropriated foreign oil interests. There is every indication that this trend will continue and further increase the pressure to expand operations on the continental shelves of the United States.

If oil import quotas were ended and the favored tax position taken away, expansion of offshore operations would probably diminish for a few years because land-based operations are considerably less expensive than offshore production. But the reduction would be only temporary, until onshore sources could no longer satisfy our needs. If we wish to slow the expansion of offshore development or to suspend it at specific places like Santa Barbara, we will only be able to do it by reducing overall oil demand and consumption. On pollution grounds there is very little reason to single out offshore production for blame. Shipment of oil by tankers continues to provide more than ten times the pollution of offshore wells. Vastly larger contributions to pollution from oil come eventually to the oceans from the end consumption and disposal of petroleum and its derivatives on land. And in the end, it is the oil transported through and disposed of in the ecosystem that is the risk, no matter where it was originally produced. In this sense, California is committing ecological suicide by her consumptive habits and "high" standard of living.

As production of oil and other commodities moves offshore, our domestic "tidelands controversy" takes on international significance. Who owns, or may take, or may profit from resources in the sea? This is no new question. Quarrels over fishery rights have sprung up intermittently for centuries and they continue today. In the case of depletable resources, the problem is especially serious: whoever removes a depletable resource removes it permanently. Even the scale of the fishing industry now requires action.

Beginning at the time of Malta's assertion of territorial rights to 200 miles offshore, the international dispute over ownership has grown. Many countries, following Malta's lead, have declared unilaterally their sovereignty over various pieces of ocean near their boundaries. The U.S. has been divided on this issue. Most industries, including oil, have supported extension of U.S. claims to the edges of the continental shelf or beyond. The official position still recognizes a 12-mile limit, pending international agreement of some kind.

In recent years there has been some sentiment for reservation of the benefits from ocean resources for men of all nations. It has been suggested that the wealth of the oceans be under the jurisdiction of the United Nations or some new world body. This seems like a humane and sensible suggestion. If we really care about mankind and the earth as we profess, we should seek to prevent private grabs of the ocean. This may be a path nearer to peace and hope for everyone.

In the meantime, so long as our addiction to energy use increases, we will have oil production. The end result will be increased oil pollution, in Santa Barbara and elsewhere, whether or not there are offshore wells and blowouts. Our system of a largely government-managed and politically powerful industry responds to assumed energy needs and defeats public interest in all other areas. We choose our future in the name of defense and nationalistic self-interest.

XII Pouring Oil on Troubled Waters

Few can be induced to labor exclusively for posterity; and none will do it enthusiastically. Posterity has done nothing for us; and theorize on it as we may, practically we shall do very little for it, unless we are made to think we are at the same time doing something for ourselves.

Abraham Lincoln

Ashes to ashes, and clay to clay, if the enemy doesn't get you your own folks may.

James Thurber

. . . the pathos of . . . Modern America: "What am I doing in all this junk?"

Jack Kerouac

The Santa Barbara oil spill was a dramatic, highly visible event. But for every Santa Barbara, for every drama-laden spill that arrests the public's attention, there are hundreds of unsung oil pollution incidents. There are several every day of the week, every week of the year, and their frequency is increasing. We do not know exactly how many there are, because we do not police every square inch of the globe. All we know with certainty is that our estimate of the frequency of oil spills is a conservative one, and the statistics we have are alarming enough.

In the United States, the Coast Guard reports all pollution events in which a federal law may have been violated. Let us look at what has been happening in recent years. During 1966, the Coast Guard reported 178 cases of oil pollution involving a vessel. A number of other cases were charged to oil refineries or storage facilities. The U.S. Corps of Engineers, however, estimates that there were over

2000 oil spills within U.S. waters in 1966, of which 40 percent came from land based facilities. In the Los Angeles-Long Beach area alone, the California State Department of Fish and Game reported 181 oil spills.

During 1967, the Coast Guard charged 211 ships with oil pollution, in addition to a large number of onshore facilities. In 1968, there were 531 reported oil pollution incidents in the coastal and navigable waters under Coast Guard surveillance.

The greatest threat is from oil tankers. Each year there are, on the average, about 500 accidents involving tankers. Not all of these result in spilled oil, but a large number of them do. With the increasingly large numbers of tankers plying the seas, with the most enormous tanker dwarfed by the blueprints for her successor even before she is launched, oil spills of the Torrey Canyon ilk are certain to increase in number and in capacity for devastation. The amount of petroleum transported across the oceans is staggering. Each year it totals about one billion tons—almost 300 billion gallons. Each year, it increases by about 4 percent. Approximately half of the world's annual production of petroleum is shipped across the seas. Through spillage or leakage, an estimated 0.1 percent of the total transport—more than a quarter of a billion gallons—is lost each year to the marine environment. This is the equivalent of almost a hundred Santa Barbara oil spills. It equals the annual marine production of petroleum-like materials from natural processes like decomposition of organic matter, natural seeps, and so on. Thus, by the careless transport of petroleum, man is roughly doubling the annual hydrocarbon input into the sea. Some of the results of this are obvious: oil-covered beaches, dead birds, fish tasting of petroleum. Other results are less obvious: gradual changes in the biota of polluted waters. Still other results, perhaps the most insidious and dangerous of all, may be yet to be discovered.

Let us look at some of the accidents that have occurred, beginning with the wreck of the Torrey Canyon, which made many people aware for the first time of the threat of oil pollution.

The Wreck of the Torrey Canyon

The Torrey Canyon was a super tanker in her day. With a deadweight of almost 66,000 tons, she was large when she was built in Newport News in 1959. Then, five years later, she was "jumboized" in Japan, increasing her deadweight tonnage by nearly a factor of two. Ironically, she was owned by the Barracuda Tanker Company of Bermuda, a subsidiary of the Union Oil Company, and was leased to the parent company. She was named for the Torrey Canyon oil field near Santa Barbara. Union Oil was proud of the tanker Torrey Canyon and of the field from which she took her name. The field had been discovered in 1889 by one of the founders of the company; it produced abundantly for many years and was still in production as its tanker namesake plied the seas. A handsome scale model of the tanker Torrey Canyon stood proudly in the entrance hall of Union Oil's Los Angeles headquarters.

The Torrey Canyon had had at least one brush with disaster before her fatal voyage. In June 1965 she ran aground close to home, while heading toward the entrance of the Los Angeles Harbor with a full cargo. That time, tugs pulled her off safely. She was not so lucky 21 months later when, near the end of a run from the Persian Gulf to Milford Haven, she tried to save half an hour by taking the treacherous eastern route past the Seven Stones Reef and instead, impaled herself on Pollard Rock. There she hung tenaciously for six weeks, resisting salvage efforts, assaults by wind and tide, and finally, repeated bombings to set her afire. In the meantime, she poured her cargo of more than 30 million gallons of Kuwaiti crude oil into the waters of the English Channel. About five and a half million gallons, more than the entire Santa Barbara oil spill, found their way to the coast of Brittany, while another four million gallons came ashore at Land's End and along the Cornish coast. We have already discussed the damage that was done both by the oil and by the cleanup procedures.

The wreck of the Torrey Canyon shocked the world. Never before had there been an oil spill of such magnitude and malevolence so close to densely populated land. It was as if people had half-known, without actually believing in their hearts, that this kind of accident could occur. But then it did occur, and people everywhere began to ask themselves, when and where will it happen next?

The Ocean Eagle and Other Oil Spills

It happened next off the coast of Puerto Rico, a year later. The Ocean Eagle carrying 5.7 million gallons of Leona crude oil, split apart at the mouth of San Juan Harbor. Like Torrey Canyon, Ocean Eagle had been jumboized by the addition of a segment to her mid-section. Although it was reported that she ran aground, some people whispered that Eagle had been broken by the synergistic action of gusty, 24-knot winds and 10-foot waves. As she split, she immediately began to discharge her cargo into the water.

The response of the American and Puerto Rican governments was instant and fairly well organized, yet we apparently had not learned much from the experience of the Torrey Canyon. Large amounts of detergents were applied at once, contrary to the advice of wildlife experts. The most striking effect of the detergent was not on the oil or the wildlife, however, but on the beaches. San Juan's sparkling beaches are as precious to Puerto Ricans as the beaches of Santa Barbara are to Californians. The detergents on the beach altered the physical properties of the sand and other particles so that when the tide washed out it carried with it not only the oil complexed with detergents, but much of the beach sand as well. Some resort hotel owners went to bed sickened by the sight of their blackened beaches and woke up to find that they had no beaches at all. The sand and the oil were both gone, swept into the sea.

A check of the larger oil spills of 1969, the year of the Santa Barbara blowout, reveals the following incidents. In January: a Washington Natural Gas Company storage tank spilled 90,000 gallons of diesel oil into Lake Washington and

Puget Sound. A pipeline break in Lima, Ohio, poured 700,000 gallons of crude oil into the Ottawa River. Union's well A-21 blew out in Santa Barbara Channel.

In February: 10,000 gallons of residual fuel oil from an unknown source washed onto the Dutch coast at the North Sea beach of Terschelling, while smaller quantities fouled other Dutch beaches. Later in the month oil and dead birds washed ashore on the east coast of Britain. The British pollution seems to have resulted from the Dutch spillage; it consisted of a similar type of oil and its timing corresponded with shifts of the wind in the North Sea. An oil spill very easily becomes an international matter. Also in February, colliding oil barges spilled 105,000 gallons of Number 2 fuel oil into the Mississippi River near Memphis.

In March: a pipeline break in Grey Bull, Wyoming sent 84,000 gallons of crude oil into the Bighorn River. While the President's Panel on Oil Spills was holding its second meeting in Los Angeles, the tanker Yukon, carrying 6.3 million gallons of oil, ripped her hull on a submerged object in Cook Inlet, 100 miles southwest of Anchorage, Alaska. Her cargo poured into the sea. In Napa, California, vandals damaged storage tanks, pouring 20,000 gallons of gasoline and 500 gallons of weed killer into the Napa River. Two wells drilled by the Mobil Company off the Louisiana coast blew out, creating a heavy slick in the Gulf waters. In Britain, oil washed ashore on the Kent coast as resort owners were making ready for the Easter holiday. The origin of this oil was obscure.

In April: the Signal Oil Company was fined $500 for permitting an onshore spill at Elwood to contaminate the Santa Barbara Channel waters. This spill of some 1300 to 1700 gallons had resulted from two human errors: one valve on a storage tank had been left open instead of closed and another was closed instead of open. A Formosan freighter collided with an oil barge under the Greater New Orleans Mississippi River Bridge. The oil barge broke in two, spilling its 375,000 gallon cargo of crude oil down the Mississippi. Chrysler Corporation's Cycleweld Plant spilled 96,000 gallons of industrial cutting oil into the Detroit River, which carried it on into Lake Erie. The tanker Hamilton Trader collided with another ship in the Irish Sea off Liverpool and the River Mersey, releasing between 170,000 and 200,000 gallons of crude.

In May: a broken line on an oil barge at Indian Harbor, Indiana, poured 10,000 gallons of Number 6 fuel oil into the Indian Harbor Canal. There was a spill of highly toxic fuel oil near the mouth of the Piscataqua River in New Hampshire. The light-weight poison spread like a flash across the water, soon permeating the beach and tidal flats and killing millions of estuarine shrimp and other small organisms that are the prime food for game fish.

In June: between 75,000 and 100,000 gallons of Number 2 fuel oil spilled from a barge unloading at the Cities Service Oil Company storage tanks in Braintree, Massachusetts. The barge Jim Hoagland ran aground after passing through the Alma dam on the upper Mississippi River. Four of the barge's ten compartments were punctured, leaking between 60,000 and 100,000 gallons of Number 2 fuel oil into the river. Some of it came ashore, forcing closing of Alma's beaches. The rest washed 20 miles downstream, killing fish along the way. Commercial fishermen, their catches fouled, were unable to sell their fish at market.

In August: the French tanker Gironde collided with the Israeli ship Harbashan, spilling 450,000 gallons of petroleum along 15 miles of Brittany's coast. Six days later refined fuel oil of uncertain origin, but probably from the Gironde spill, washed ashore near St. Brieuc at the height of the tourist season.

In September: between midnight and 2:00 a.m., the barge Florida ran aground off Chappaquoit Point in Buzzards Bay, Massachusetts. More than 100,000 gallons of Number 2 diesel fuel leaked from her damaged tanks.

The year between February 1970 and February 1971 witnessed three spectacular oil spills, including two offshore oil well blowouts. In February 1970, an offshore well drilled from Chevron's platform "Charlie," 75 miles southeast of New Orleans, suffered a blowout and fire which ultimately involved twelve wells. Dangerous and terrifying as the fire was, it did minimize the oil pollution while it burned. After the fire was extinguished, a massive slick covering 52 square miles bloomed from what had been "Charlie."

On December 1, 1970, Shell's well number B-21 blew out in the Gulf of Mexico. Almost immediately, escaping gas was ignited and soon seven wells were involved in the blowout fire. Two members of the platform crew drowned after they jumped into the Gulf to escape the flames. A third man was missing and presumed drowned. A fourth died from injuries and burns. Altogether, about 40 men were injured, some critically. Fires roared uncontrolled throughout the month, while relief wells were drilled to cut off the flow of gas and oil into B-21.

While the Shell fire was lighting up the coast of Louisiana, the Navy, following its long-standing custom, dumped a load of waste fuel oil off the coast of Florida. This oil created a slick covering 760 square miles and threatening 50 miles of Florida beaches. The Navy was embarrassed. The winds were not supposed to have washed the oil toward shore. The Navy dumps oil into the sea regularly. It is immune to the law, because federal laws do not apply to the government itself.

On Monday, January 18, 1971, disaster once again visited the coast of California. Shrouded by an all-concealing fog, the tanker Oregon Standard rammed her sister ship Arizona Standard under the Golden Gate Bridge. Wedged together and leaking their cargoes of bunker oil in a steady stream, the tankers, a Siamese twin of a wreck, were swept by the tide under the bridge and into San Francisco Bay. There the decision to separate the conjoined twins brought about disaster. What had been a manageable stream of escaping oil erupted to a violent outpouring of more than a million gallons as the tankers were broken apart. Changing tides spread the oil around the bay and up and down the coast as, still cloaked by the fog, the oil stealthily did its work before an unsuspecting public could see what was going on.

Growth as a Disease

In the United States we constantly recite to ourselves and to each other a series of desires and expectations which are in themselves mutually contradictory. We wish for drugs that are absolutely safe, we hope for pest and weed control with great effectiveness but no side effects, and we demand that national security deliver us safely from any threat without the security preparations themselves constituting a threat. Combined with all this is our almost fanatical belief in continuous growth. We take pride in the fact that our energy supplies have doubled in each of the last three or four fifteen-year periods. We *used* to express considerable pride in our population growth, although we are now dimly aware that this process must stop somewhere. Underlying the unstated American dream is a notion of an endless continuation of growth, which provides for everyone a future that is not only infinitely long but cheap, safe, and delightful as well.

In the long run, all of these hopes are illusory. We are all doomed. An individual death faces each of us, and the situation for the future of the planet, although the time is longer, is no better. We are on a smallish planet, 93,000,000 miles from a middle-aged star. At some time, perhaps five or six billion years from now, this star will expand and heat up the earth, rendering it uninhabitable. This future, because of its remoteness, does not concern us very much, although small children often become worried when they first learn of it.

At shorter range, the sweep of evolutionary history for the last 500,000,000 years or so promises with as much certainty as we can ever have that our species, *Homo sapiens,* will disappear from the earth as the inexorable course of evolution and environmental change produces species that are better adapted to the current conditions. We have already lasted quite a long time for a mammal of our size and complexity. So we may be extinct in another million years or so, but this future too seems remote and unreal.

Though cosmic and evolutionary extinction of life through causes beyond our control are too remote for concern, our species is still in danger. For the last one or two centuries, population, energy consumption, and almost everything else we look at have been increasing at an exponential rate. Although this growth explosion is recent in terms of all human history, even the oldest among us has his experience limited to the period of exponential growth. In these circumstances, we tend to take our present situation as the normal one and to look askance at those we call "doomsayers" who say that it cannot continue. Yet we *know* that exponential growth cannot continue because it leads to projections which will either exceed the physical limitations of our planet or result in other kinds of absurdities within the easily foreseeable future. Continuing our present population growth rate, for example, would lead to a population of 30,000,000,000 people in about 2075–a mere hundred years from now. Continuing such a growth rate much past the year 2400 would lead to a population density equal to that of downtown Manhattan over the entire surface of the earth including the oceans. Such a prediction is absurd and we can be quite sure it will not come to pass, yet in the short run we behave as if that is exactly what we expect. Opinions vary on how far into the future we should try to peer, fuzzy as such glimpses and guesses are; but our grandchildren will thank us if we at least look to a future of a thousand years or so, rather than to the ephemeral period of our own lifetime.

If growth is the disease of our species, one of its possible outcomes is death. It is much more likely, however, that nature, which deals regularly with growth explosions among the lemmings and bacteria, can put an end to human expansion as well. We wish to take a hand in our own future to avoid the disease, starvation, misery, and war that are implied by a reliance on natural limitations. We already have a great plenty of these unlovely commodities.

We could take some immediate steps toward limiting the expansion of energy usage within the framework of our present institutions. If, for example, the structure of electrical rates were reversed so that the first units were cheapest and subsequent units increasingly costly, industry and most individuals would respond by regarding the marginal costs of electricity carefully. The degree of rate increase with additional usage could, in principle, be adjusted to achieve any desired reduction in usage. Such a measure is neither regressive nor repressive. Similar plans could be designed to cut down energy use by automobiles. But these are interim procedures, to buy time for real change. Steady-state energy requirements in the long run will necessitate both a steady-state population and changes in life style and the institutions of our society.

Institutions and Policies

Our best dreams for our society envision people with individual rights in a peaceful nation. Government and industry are but instrumentalities established by people in their attempts to realize these dreams. In turn, the body of law, common agreement, and organizational and administrative procedure by which the institutions of society operate are the elaborations of these institutions. We have suggested in earlier chapters that some of this machinery is not working properly. Institutions should be subject to constant scrutiny and criticism and should *always* be subject to change. It will not do to erect rigid mythologies about our society and talk of "rights"—rights of government and industry, where no such rights exist.

Secrecy must be fought at every turn. The tendency of government toward secrecy, notorious in the *Pentagon Papers,* is compounded by the complexity of government today. Before accusing those bad guys with their evil motives again, let us note the fact that the federal government now spends nearly $500 million annually on information systems designed to keep one part of itself informed on the research and development projects of another. Most of these systems are only marginally successful. The blocking of free information flow by vested interests and political pressures cannot be tolerated, but it will take more than good will to end secrecy.

If we have an open government there will always be criticism. On any issue, public opinion polls have one message: there is always a substantial minority, usually 20 percent or more, dissenting from any majority view. If an election is won by more than a few percent, it is a "landslide." This leads to a stubborn problem in our kind of government: how to have majority rule without oppressing minorities. The minority may be the people of a particular area, like Santa Barbara. The minority may be the people who would give up their automobiles to save the birds. We have not been able to solve this kind of conflict very successfully.

And the free enterprise myth goes on. We have shown that groupings of industry like oil and steel have strayed far from the traditional free enterprise system. They are strongly and probably permanently enmeshed with the government at various levels and in many ways. Only when we admit to having a mixed economy, with the government involved ever more deeply in its management and regulation, will we have a chance of treating the actual problems.

We should separate, so far as possible, the regulatory machinery from the administrative machinery. One cannot promote policies and operate in the quasi-judicial role of regulator at the same time. The courts have shown their inability to resolve these problems. Perhaps we need a new, fourth branch of government to handle regulatory matters. Perhaps only a more complete reorganization of the government can do the job.

But organization alone will not solve all the problems. At any one time, a great many different kinds of activities and problems occupy our government. There are probably as many optimal organizational plans as there are problems and activities, but we can operate with only one organizational structure at a time. This problem already seems intractable and is growing worse. The only solution so far suggested is decentralization of authority wherever possible, so that regions or groups may organize better for their particular needs.

In the end, the government is only people, and its success or failure is a kind of summation of the successes and failures of its civil servants and other officials. The matter of technical expertise was at considerable issue in Santa Barbara. Past practices have left the civil servants at some technical disadvantage to industry—and these civil servants are often suspected of self-serving and political machinations besides. If we want experts in government we will have to pay for them, as industry now does. If the Civil Service is still unable to get trustworthy technical people, we must change the Civil Service. We might place fewer—but better and more highly paid— technical people in positions of authority in government, and do this without regard for their political opinions. The advantages and dangers of this are illustrated by one example in our current situation. The inner circle of the Scientific Establishment numbers no more than 300 or 400 men who have lasted longer in their influence on government than anyone except the judges of the Supreme Court. Since World War II, essentially the same men have populated the inner technical circles of government. They are honorable and well-intentioned men, but there is no reason to believe that the aspirations of the high subculture of scientists have much in common with the aspirations of the majority of Americans. We must use technical groups within the government only for their expertise on technical matters, relying on conflict between them to point out the problems that remain unsolved.

It is usually complexity that defeats us: complexity of government, of technical matters, of the profusion of documents and information, and of the very division of authority to do anything. No citizen can expect redress for grievances unless he knows where to look for it. In Santa Barbara, many citizens seemed to expect redress from quarters from which it could not come. This is the prescription for useless conflict and frustration. Would an adversary government agency help? One funded at a level of perhaps 1 or 2 percent of the national budget, whose sole function would be to search for redress for citizens?

Society and the Future

Our problems come upon us gradually. Only the amount of attention we pay to them fluctuates. The dramatic events of the Santa Barbara oil spill drew attention to the environment, with its range of creeping problems. If, as it now appears, the spill was not a disaster for the environment, we did find a formerly abundant population of animals diminished by half in little more than ten years and a group of worms reduced 20 percent in number of species—not as a result of the oil spill but because of general increase in pollution and human activity. These are threatening messages from the natural world. When diversity diminishes, we are all poorer, and the web of life is threatened. How imminent such threats are we do not know. Yet every year we lose the equivalent of 100 Santa Barbara oil spills into the environment before even reaching the end uses of petroleum products. Much, much more is supplied to water and air by "final" disposal.

It is clear that our social and cultural arrangements have resulted in the gradual growth of our environmental problems and that our lifestyle, including the addiction to growth, promises more and more serious problems. Conflicts of interest representing real dilemmas result, and the government, meant to resolve such conflicts peacefully, has been increasingly unable to do so. In Santa Barbara, for example, banning oil production in the channel does not address the problems to be expected if a growing percentage of the area income and tax base is removed, and at a time when the nation faces a growing energy shortage—whatever one may think of the reality of our energy "need." Banning offshore oil would mean social disruption at a human as well as an economic price. But continuing growth of the oil industry brings its own price, as we have seen.

The government does not fail because it is weak, or, as radical rhetoric will have it, decadent. It is strong and, at times, threatening. But it is badly suited to resolve our present needs and conflicts. The first problem is setting goals in a diverse society. In a diverse society, this process is of necessity accompanied by conflict. We should not try to reduce diversity, for here, as in natural ecological communities, lies the strength and hope for the future. Instead, we must welcome and promote open discussion and attempt to keep the conflict short of actual violence. Goals should always be provisional and open to change. In any case, sloganeering and mythology derived from past successes and failures are not likely to be helpful.

Sharing the world with a bewildering variety of societies is difficult; sharing the world with the hundreds of thousands of other species of living creatures is still more difficult. But if we do not share it we will lose it. Man is not especially complex in comparison to other animals. It is his elaborate forebrain that causes him to search for meaning and satisfaction as well as survival. Even with this brain of his, it is not clear that the hairless great ape called man can modify his behavior before natural limitations do it for him. But his brain has been the secret of his survival so far, and he still hopes to survive.

Notes and References

NOTES

Chapter II

1. *Santa Barbara: A Guide to the Channel City and Its Environs* (New York: Hastings House, 1941), pp. 25-55. Much of the information in this section was obtained from this book, written by the Southern California Writers' Project of the Works Project Administration.
2. As quoted in Union Oil Company of California Research Library Pamphlet, I.D. No. 6-6-62-46, "Oil Seep Literature Research: Santa Barbara Channel Area" (1969).
3. As quoted in U.S.G.S. Professional Paper 679, "Geology, Petroleum Development, and Seismicity of the Santa Barbara Channel Region, California" (U.S. Gov. Printing Office, 1969), p. 14.
4. *Ibid.*, p. 14.
5. As quoted in Union Research Pamphlet No. 6-6-62-46.

Chapter III

1. Second Report of the President's Panel on Oil Spills (U.S. G.P.O., 1969), p. 3.

Chapter IV

1. As quoted in W. H. Hutchinson, *Oil, Land, and Politics,* vol. 1 (Norman: University of Oklahoma Press, 1965), p. 58.
2. *Ibid.*, p. 58.
3. Pamphlet, California Petroleum Company, "A Description of the recently Discovered Petroleum Region in California. With a Report on the Same by Professor Silliman" (New York: Francis & Loutrel, Printers, 1865), p. 20.
4. Much of the information in this and the following sections was obtained from E. R. Bartley, *The Tidelands Oil Controversy* (Austin: University of Texas Press, 1953).
5. *Offshore-Onshore Petroleum Study, Santa Barbara Channel,* by the Oil Well Inspection Dept. (1967).

Chapter V

1. Santa Barbara *News-Press*, Feb. 20, 1969, p. A-8.

Chapter VI

1. U.S.G.S. Professional Paper 679 (U.S. G.P.O., 1969), pp. 43-44.

Chapter X

1. Walmsley, David J., *Oil Pollution Problems Arising Out of Exploitation of the Continental Shelf: The Santa Barbara Disaster* (1969). Report prepared for the Harvard Law School.

Chapter XI

1. Wright, M. A., "U.S. Energy Crisis and What Can Be Done About It," *Ocean Industry 6,* No. 6 (1971), pp. 11-18.
2. Chamberlain, Edward, *Theory of Monopolistic Competition* (Cambridge, Harvard University Press, 1950).

REFERENCES

Chapter I

"National Multi-Agency Oil and Hazardous Materials Pollution Contingency Plan" (September 1968).

The Santa Barbara *News-Press,* the *Los Angeles Times,* and other California newspapers published oil spill stories throughout all of 1969, and provide a good starting place for further reading.

Chapter II

General background material may be found in U.S.G.S. Professional Paper 679 (U.S. G.P.O., 1969).

Chapter III

Let's Ban Blowouts. Reprints of 4 articles from *Drilling Magazine,* published by Drilling Magazine (1967).

Offshore-Onshore Petroleum Study, Santa Barbara Channel. Phase I: Effects of Federal Leasing, Outer Continental Shelf (November 1967), prepared by the Oil Well Inspection Dept. for the Board of Supervisors, Santa Barbara Co., Cal.

Plain Facts About Oil, U.S. Dept. of the Interior, Office of Oil and Gas (U.S. G.P.O., 1963).

Chapter IV

Nash, Gerald D., *United States Oil Policy 1890-1964* (Pittsburgh: University of Pittsburgh Press, 1968).

Robinson, W. W., *Land in California* (Berkeley: University of California Press, 1948).

White, Gerald T., *Scientists in Conflict: The Beginnings of the Oil Industry in California* (San Marino: Huntington Library Publications, 1968).

Chapter V

References to most of the topics discussed in this chapter may be found in the Santa Barbara *News-Press* or the *Los Angeles Times.*

Chapter VI

The Oil Spill Problem. First Report of the President's Panel on Oil Spills. Executive Office of the President, Office of Science and Technology (U.S. G.P.O., 1969).

Offshore Mineral Resources, A Challenge and an Opportunity. Second Report of the President's Panel on Oil Spills. Executive Office of the President, Office of Science and Technology (U.S. G.P.O., 1969).

Chapter VII

Report of Special Panel on the Future of the Union Oil Lease, Executive Office of the President, O.S.T. (1969).

Chapter VIII

"Event Information Reports," Smithsonian Institution, Center for Short-lived Phenomena (Cambridge, 1969).

Jones, L. G., Mitchell, C. T., Anderson, E. K. and North, W. J., "Just How Serious Was the Santa Barbara Oil Spill?" *Ocean Industry 4,* No. 6 (1969), pp. 53-56.

Klaus, R. L., "In the Case of Santa Barbara," *Our Sun 34,* No. 2 (1969), pp. 3-17.

Neushul, Michael, *Santa Barbara Oil Pollution, 1969: Final Report Dealing with the Early Stages of the Santa Barbara Oil Spill.* U.S. Dept. of the Interior, F.W.P.C.A. Contract No. 14-12-516 (1969).

Chapter IX

Bates, Martha, *Water Pollution by Oil Spillage: A Literature Search* (Southern Methodist University, Industrial Information Services Science Library, 1969).

Biological and Oceanographical Survey of the Santa Barbara Channel Oil Spill 1969-1970, in 2 vols. (University of Southern California, Allan Hancock Foundation, 1971).

Blumer, Max, Sanders, H. L., Grassle, J. F. and Hampson, G. R., "A Small Oil Spill," *Environment 13,* No. 2 (1971), pp. 2-12.

Buchsbaum, Ralph, *Animals Without Backbones,* rev. ed. (Chicago: The University of Chicago Press, 1960).

Oil Spillage Study Literature Search and Critical Evaluation for Selection of Promising Techniques to Control and Prevent Damage. (Richland, Washington, Battelle Memorial Institute, 1967).

Ricketts, E. F. and Calvin, Jack, *Between Pacific Tides,* 4th ed., rev. by Joel Hedgpeth (Stanford: Stanford University Press, 1968).

Chapter X

Dedera, Don, "Santa Barbara & Beyond," *Oceans Magazine 3,* No. 3 (1970), pp. 17-32.

Hearings on H. R. 6495, 6609, 6794, 7325; 91st Cong., 1st sess. (1969).

Hearings on S. 1219, 2516, 3351, 3516, 4017, 3093; 91st Cong., 1st. sess. (1969).

Henkin, Louis, *Law for the Sea's Mineral Resources,* Columbia University, Inst. for the Study of Science in Human Affairs Monograph No. 1 (1968).

Chapter XII

"Horizon to Horizon," *Environment 13,* No. 2 (1971), pp. 13-21. A staff report on some recent oil spills.

Landsberg, H. and Schurr, S., *Energy in the United States: Sources, Uses, and Policy Issues* (New York: Random House, 1967).